MALE HOMOSEXUALITY IN SOUTH AFRICA

Identity formation, culture, and crisis

Gordon Isaacs
and
Brian McKendrick

1992
Cape Town
Oxford University Press

OXFORD UNIVERSITY PRESS

Walton Street, Oxford OX2 6DP, United Kingdom

Oxford New York Toronto
Delhi Bombay Calcutta Madras Karachi
Petaling Jaya Singapore Hong Kong Tokyo
Nairobi Dar es Salaam Cape Town
Melbourne Auckland

and associated companies in
Berlin Ibadan

Male Homosexuality in South Africa
Identity formation, culture, and crisis
ISBN 0 19 570715 X

First published 1992

© Oxford University Press 1992

OXFORD is a trademark of Oxford University Press

All rights reserved. No part of this publication may be reproduced, stored in a retrieval system, or transmitted in any form or by any means, electronic, mechanical, photocopying, recording or otherwise, without the prior written permission of the copyright owner.

Within the U.K., exceptions are allowed in respect of any fair dealing for the purpose of research or private study, or criticism or review, as permitted under the Copyright, Designs and Patents Act, 1988, or in the case of reprographic reproduction in accordance with the terms of licences issued by the Copyright Licensing Agency. Enquiries concerning reproduction outside these terms and in other countries should be sent to the Rights Department, Oxford University Press, at the address below.

Published by Oxford University Press Southern Africa, Harrington House, Barrack Street, Cape Town, 8001, South Africa

Set in 10 on 12 pt Garamond by Theiner Typesetting (Pty) Ltd
Cover reproduction by Fotoplate
Printed and bound by Clyson, Cape Town

Contents

Dedication		IV
Acknowledgements		V
Preface		VI
Introduction		X
Chapter 1	Homosexual identity growth	1
Chapter 2	Crisis and the growth of homosexual identity	38
Chapter 3	The homosexual sub-culture and homosexual identity	65
Chapter 4	The anatomy of a homosexual sub-culture: the Greater Cape Town area	89
Chapter 5	AIDS: the new homosexual crisis	112
Chapter 6	Development and nature of the formal gay movement in South Africa	138
Chapter 7	Homosexual identity formation, culture, and crisis: an empirical study	165
Chapter 8	Conclusions and implications for helping professionals	206
Appendix A	*A case study of a person with AIDS: Robert*	233
Appendix B	*Survey questionnaire*	241
Appendix C	*Glossary of gay vernacular in use in South Africa*	247
Bibliography		251
Index		276

This book is dedicated to all those
people living with HIV and AIDS.

Acknowledgements

This book is based in part upon a Ph.D. thesis entitled 'The Growth of Homosexual Identity: An Empirical Study from a Social Work Perspective of Crisis in Sexual Identity Experienced by a Sample of Homosexual Persons in Cape Town, With Special Reference to Cultural Factors', submitted by *G M Isaacs* to the University of Cape Town.

Grateful thanks are acknowledged, in the first instance, to Emeritus Professor *Brunhilde Helm* for her sensitive guidance, constructive opinions, and support at all times.

Special appreciation is extended to the following people:

Pat Halford, a secretary in the School of Social Work, for her administrative assistance;

Hans Normann and *Digby Warren* for their technical and research assistance; and finally

To the members of GASA 60-10 for their generous and willing cooperation.

Preface

'It is a capital mistake,' says Arthur Conan Doyle's famous character, 'to theorize before one has data.' This empirical route had been indicated by Aristotle, but for a long time progress along it was slow. The past century has, however, seen rapid acceleration in the employ of methods we now associate with 'science'. The 'natural' sciences took the lead, with the 'social' sciences following fast behind; although many enquirers are far less comfortable observing themselves or their fellow humans than observing the 'external' universe.

In the nineteenth century, Alexander Pope warned us that the proper study of humanity is people; in his time he was not really taken at his word. In particular, several aspects of human behaviour received no systematic attention from the data-gatherers. Sexuality, hemmed in as it is by restraint if not taboo, is a primary example. We have had to wait until the middle of the present century for a Kinsey.

If sexuality in general has been difficult to study empirically, homosexuality has been even more so. Broadly speaking, Western culture has acted as if homosexuality were a human condition that everybody knew about but that didn't really exist. When, in recent decades, attention was reluctantly persuaded to it, perplexing and indeed conflicting theorizing proliferated — the theorizing without data that is Doyle's 'capital mistake' and which Isaacs and McKendrick in this book eschew.

As these two authors illustrate in their extensive reviews of published material, clear thinking about homosexuality remains rare (although all the more welcome for that). Current terminology has hot helped. A *homosexual* is usually thought of as male (as if women never are sources of sexual attraction to other women). Sometimes the prefix in the word *homosexuality* is confused with a Latin word, as in homo sapiens ('homo', says Shakespeare, 'is a common name to all men'), whereas the actual root is Greek, meaning 'same'. Confusion can only be further confounded by the use of *heterosexuality* in opposition to *homosexuality*. *Hetero* bears at least two connotations, themselves sometimes combined. As a Greek root, it means 'the other of two', but (as in *heterogeneous*) it also signifies difference. In other instances it might connote diversity, irregularity, even (as in *heterodoxy*) deviation. Increasing use nowadays of the old word 'gay' might for some clear the air; but, as the authors of this book point out, not all homosexual behaviour can rightly be attributed to gays.

Whether or not the reader of this book belongs to one of the several professions to whom men and women increasingly are turning for help in their relations with others, he/she will find the authors threading sure-footedly through what, without their guidance, might seem a maze. In addition, they place homosexuality in a framework which stresses similarity rather than difference and commonality rather than idiosyncrasy; a framework, that is, of a society shared by all its members. Homosexuality is not a condition whose manifestations will be clinically identical regardless of where it is found, such as malaria. It expresses itself within the prevailing culture, and according to that culture, just as other forms of interpersonal behaviour do.

If the behaviour patterns of a homosexual individual are best understood by referring to the culture with which he/she identifies, this (in part at least) is because of the attitudes towards homosexuality that that culture conveys. In most Western countries homosexuals experience attitudes towards them that are essentially characteristic of discrimination and stigmatization in general. Homosexuals share much of the experience of other minorities.

South Africa is a country, like others with a colonial history, that has not been kind to minorities. Moreover, it has clung to its oppressions when other former colonies (in word at least, if not in deed) have sought to do away with theirs. The present book is a book out of Africa. The first author is from Cape Town, the second from Johannesburg; their book is published in South Africa. That the book has proved to be possible is in itself an indication of major social change, yet this should not detract from the credit due to the authors for their courage and determination in undertaking the work they have done in such a sensitive field.

The two authors have for many years been colleagues. Dr Gordon Isaacs at the University of Cape Town and Dr Brian McKendrick at the University of the Witwatersrand (Johannesburg), respectively, at present head two of the major Schools of Social Work in South Africa. Both universities represent a tradition of liberal thought and open enquiry, thus setting the stage for books like the present one, a pioneer in its field in South Africa. The next step, as the authors would agree, is to carry out data-based studies of homosexuality within the African and other cultures in South Africa less 'Western' than the authors' respondents.

Social workers, physicians, psychiatrists, psychologists, and other counsellors might fruitfully explore the connections the authors establish between crisis intervention theory and practice on the one hand and, on the other, the traumas often associated with the homosexual's public acknowledgements of his homosexual identity. Homosexuals, themselves, seeking to lead balanced and rewarding lives within social environments that remain largely hostile, will find that the book provides a rich source of insights. The idea that their generally comfortable and comforting 'gay sub-culture' might for gay people carry its own inherent dangers could be revelatory.

The Reverend John Donne, in the early seventeenth century, shortly after his marriage, lies abed in London with an epidemic illness diagnosed then as 'relapsing fever'. Elizabethan and Jacobean remedies do not much avail (dead pigeons, for instance, at his feet!). He lies sleepless, while in the streets outside the bells toll, again and again, for the dead victims of the same disease being carried to their burials. The sick man has been writing his *Sermons* for publication. He pens what has become one of the most famous prose passages in all the English language:

> No man is an island, intire of itselfe; every man is a piece of the continent, a part of the main ...

A piece of the continent, a part of the main. Social change in general is slow-paced, but this is the way homosexuals are coming to perceive themselves; even more gradually do others perceive them. A piece of the continent, an enriching part of the main.

This book has important findings about the crises associated with what homosexuals call 'coming out'. As ostracism, prejudice, and hostility wane, so too should 'coming out' become less traumatic. In due course, homosexuals might face no more than the stresses encountered by humans in general in establishing acceptable and rewarding expressions of their sexuality; no easy task in any case, but part of the main.

Brunhilde Helm
Professor Emeritus, University of Cape Town
November 1991

Introduction

This book is about a South African minority group — male homosexuals — in a country where prejudice towards minorities has become a way of life. While other disadvantaged groups, such as black South Africans and women, have had or are in the process of having their human rights acknowledged and legal discrimination against them removed, homosexuals remain a socially unaccepted group. They are liable to legal sanction for the expression of their sexual preferences, subject to religious condemnation, discriminated against in many work settings, and, as this book will show, they themselves often become active agents of their own oppression.

They are a substantial minority group, since an estimated one out of 10 South Africans has a homosexual identity, even if this identity is disguised, denied, or suppressed. The three to four million South Africans of fundamentally homosexual orientation, half of whom are males, are usually overlooked in contemporary debates about human rights, as their circumstances are overshadowed by the demands of more prominent groupings of currently disadvantaged people.

Hence, while homosexuals in many other parts of the Western world (such as North America, Britain, Western European countries, and in some Australian states) have won legal freedom and civic acceptance as citizens in good standing, South African homosexuals remain locked in to the status of social deviants and potential or actual criminals.

This book seeks to explore how some males acquire a homosexual identity, and the consequences which this identity carries for them in South Africa as they attempt to deal with their life roles and transitions. The book's foundation is the view that homosexual identity, like heterosexual identity, is not pathological, but one of a number of sexual identity alternatives to which an individual is irrevocably inclined by his personality and life experiences.

However, being homosexual in a predominantly hostile heterosexual world has many consequences. One is that developmental and life transitions that would otherwise have been well within the person's coping capacities become inordinately stressful and demanding, leading to crises that disable and immobilize the individual. Hence, a second foundation of this book is the concept of crisis and its part in homosexual identity formation.

Since homosexuals do not obtain social support from the major South African culture which rejects their identity as invalid and unacceptable, a homosexual sub-culture has evolved which validates the homosexual identity, and at the same time entraps the individual in a sub-cultural milieu that stresses 'difference' and 'separateness', contributing further to identity crisis.

Thus homosexual identity formation, culture, and crisis are the three interrelated themes of the book. The book's primary purpose is to promote understanding of how homosexual identity develops, and its consequences for the person. Its secondary purpose is to use this knowledge and insight to suggest appropriate interventions by helping professionals — social workers, psychologists, medical doctors, and psychiatrists — that are directed towards strengthening the ability of homosexuals to take charge of their lives with dignity, to their own advantage and to that of the South African society of which they are a part.

The sources of material contained in this book are multiple. Use has been made of the scientific literature on homosexuality and homosexual identity formation, both South African and international; observation and analysis of homosexual behaviour and sub-cultural structures and organizations; and the findings of an empirical research study conducted in Cape Town, in which the respondents were members of a homosexual organization. In addition, frequent use has been made of extracts from records of the writers' clinical practice with homosexual clients.

The book has been organized into eight chapters, each dealing with a specific component of the subject. The first chapter explores and analyses themes from the literature in respect of homosexual identity growth as a dynamic process, and in this context the writers present their own model of homosexual identity formation.

Chapter 2 links homosexual identity growth to crisis, and attention is directed to a model of crisis intervention for use in response to homosexual identity crises.

Chapters 3 and 4 focus on the homosexual sub-culture, and sub-cultural influences on identity formation. This issue is dealt with generally in Chapter 3, while Chapter 4 homes in specifically on the homosexual sub-culture in one region of South Africa, the Greater Cape Town area. The anatomy of the sub-culture is analysed, and comment is offered on how and why the Cape Town sub-culture is both similar to and different from the sub-cultural scenes in other major South African centres, especially Johannesburg.

Chapter 5 is devoted to the major crisis of HIV infection and AIDS for homosexuals and the homosexual community. HIV issues and AIDS-related crises are identified and explored, and the model of crisis intervention presented in Chapter 2 is illustrated in detail in regard to AIDS and issues associated with it. A major case example (presented in Appendix A) and a number of clinical vignettes are used to support the discussion.

Chapter 6 is concerned with the development and present nature of the formal gay movement in South Africa. The international literature, particularly that from North America, consistently emphasizes the rise of an activist gay liberation movement as a critical force in establishing the validity of a gay identity. An analysis is made of why South Africa does not have a unified, national, and prominent gay liberation movement, and the consequences of this for identity formation and homosexual crises.

The first six chapters establish the writers' central assumptions about homosexual identity formation, culture, and crisis. In Chapter 7, these assumptions are drawn together and restated, and are then empirically tested through a research study involving homosexual men in the Greater Cape Town region. The research study is described, and findings are presented and discussed in the light of theoretical constructs and the findings of other research studies.

Chapter 8, the final chapter, presents the 10 major conclusions of the book. These are discussed in relation to the findings of the research study, and the theory, observational analysis, and clinical material presented in preceding chapters. Finally, the implications of the findings for effective and appropriate intervention by helping professionals are identified.

In the course of the book five principal concepts recur. These are defined below:

▶ *Crisis* occurs when a person faces an obstacle to important life goals that is, for a time, insurmountable through the utilization of his customary methods of problem-solving (Caplan, 1964:44).

▶ *Homosexuality* is a broad spectrum of psychological, emotional, and sexual variables in a state of interplay between people of the same sex. Homosexuality is not only sexual attraction between people of the same sex, but also includes an emotional as well as a physical bond; a fantasy system; and elements of symbolism, eroticism, and sexuality. Homosexuality can be experienced in different degrees (adapted from Isaacs and Miller, 1985:327).

▶ *Gay* is a term that predates the term 'homosexuality' by several centuries, and has generally been employed with far greater precision: it is normally used to describe people who acknowledge an erotic preference for their own gender. (In a prison, for example, many people may be involved in homosexual acts or even relationships without thinking of themselves as 'gay'.) 'Gay', says Boswell (1980:44) 'refers to persons who are conscious of their erotic preference for persons of their own gender as a distinguishing characteristic or, loosely, to things associated with such people, [like] "gay poetry".' 'Gay' may also refer to a state of feeling guilt-free about sexual preferences, or freedom from homosexual oppression, as well as being able to identify with and participate in a sub-culture. Not all homosexuals are 'gay'.

Nevertheless, throughout this book the terms 'gay' and 'homosexual' are used interchangeably, although, when appropriate, direct reference is made to the political and psychological impact of 'gay' versus 'homosexual'.

▶ *Gay community* as a term is really a misnomer (Milligan, 1975). No sense of true gay 'community' exists in South Africa. Community suggests geographical location, a set of interconnected systems, structural arrangements for survival and adaptation, a development of interactive relationships, and shared ways of thinking, feeling, and acting, all of which are internalized by the whole population, and with which each individual identifies himself to a particular degree according to his personal living experience (adapted from Ferrinho, 1981:4). In South Africa, the 'gay community' is a colloquial expression symbolizing a *collective of people* who share a common sexual identity. Furthermore, the concept 'community' in fact pertains only to a minority of homosexuals, in essence those who are part of the gay sub-culture, and who endorse and subscribe to its ethos.

▶ A *sub-culture*, according to Bronski, is

any group excluded from the dominant culture, either by self-definition or ostracism. The outsider status allows the development of a distinct culture, based, however, on the mainstream. Over time, this culture creates and recreates itself — politically and artistically — along with, as well as in reaction to, the prevailing cultural norms. No counterculture [*sic*] can define itself independently of the dominant culture. By definition it is distinct, yet there is always the urge, if only for survival's sake, to seek acceptance (Bronski, 1984:6).

Other key concepts are defined when they arise in the text. In addition, from time to time use is made in the text of homosexual argot. These latter terms are defined in the Glossary of Gay Vernacular in Use in South Africa provided as Appendix C of this book.

1 Homosexual identity growth

Introduction

Overlooking the diversity of its nature, sexual identity has usually been conceptualized as if it were no more than a simple dichotomy: a person is either heterosexual ('straight') or homosexual ('gay'). Yet, at the same time, sexual identity has been puzzled over in the thought laboratories of academic alchemists for centuries. Scientific discourse, occurring chiefly in European and American medicine, began to probe more deeply into the oversimplified 'either/or' version of sexual identity. The architect of this movement was Benkert, a Hungarian physician who coined the term *homosexuality*, and who provided the first critical explanation of sexual behaviour. Benkert's response was chiefly due to the alteration of sexual laws and codes in the Napoleonic era, whereby same sex behaviour was placed on an equal basis with that of opposite sex behaviour.

Following upon Benkert, Ulrich, a German philosopher, coined the term *Uranian* ('Urning' in German), which embodied the concept of a third sex — in essence, a woman's mind in a man's body. This notion was the precursor of the series of *bio-psychological* theories formulated by scholars such as Freud, Krafft-Ebing, Moll, and Hirschfield. The contribution of these theorists towards the understanding of homosexuality included the assumptions that homosexual behaviour was genetically determined, or that it was a form of arrested development; that homosexuality was a congenital disease;

that hormones, genes, and sexual instincts lodged in the cerebral cortex were transmitted to the foetal brain; and that homosexuality was an inversion, a process which, if not overcome, would result in bio-sexual degeneration.

The discourses which produced the above assumptions were attempts at scientific truth, but they also reflected moral judgements and a belief that people's biological gender is the reality upon which all human relationships are dependent. Concern about this form of fundamentalism is described by De Cecco and Shively (1984b:13), who sum up the bio-psychological theorists in the following words: 'By portraying homosexuals as "intersexuals", unfinished exemplars of their biological sexes, they were assigning them to an inferior rank on the scale of social values, while reassuring the prevailing role stereotypes.'

Sexual expression and interaction have often been described according to the premises of biological determinism. Even with the onset of enlightenment, enhanced by the post-Freudian renaissance, a bio-genetic framework has governed the focus of enquiry. The innovative work of Kinsey and his associates (1948) did little to dispel the indelible damage of the powerful medical model. With the noticeable escalation of homosexual literature, particularly from the mid-1970s to date, empirical studies have contradicted the beliefs of past eras. Recent research reveals a far broader range of opinions without necessarily denigrating homosexuality to the state of an illness. Studies such as those of Bell and Weinberg (1978), Coleman (1982), Hoult (1984), Masters and Johnson (1979), Minton and McDonald (1984), and Spada (1979) collectively reflect and examine the diversity of sexual relationships among men and women, and highlight gross shortcomings within past understandings of gender development, the nature and intent of homosexual sex patterns, homosexual identity, bio-genetic disputes, and homosexual lifestyles. In essence, the key issue to emerge from these more recent studies is that homosexual behaviour reflects feelings, attitudes, and sexual expressions within the context of relationships, which are held by the individuals alone, or in conjunction with others.

In overviewing the literature, both past and recent, a sense of disquiet is engendered — a result of the explosion of varied and often conflicting academic and literary forays into the nature of homosexuality. Weeks (1977:33) captures this when he concludes that 'Homosexual behaviour cannot be crammed into any one predetermined mould, because it [homosexuality] pervades many different aspects of social experience'.

Concern can indeed be voiced over the mushroom-like cloud of contradictory thinking about 'homosexuality' that hangs over the heads of people who wish to understand the phenomenon, and who have to sift through diverse and opposing opinions in order to establish a sense of verisimilitude for themselves. Listed below are four pointers which can guide the reader into the heart of this often confusing debate:

▶ Homophile writing has overlooked some basic literature that is relevant to human identity generally. A specific example is the work of Erik Erikson who, as a post-Freudian and an ego-psychologist, has offered invaluable theoretical treatises on stages of identity development. Yet nowhere in the excellent book on homosexual identity development published by the Haworth Press in 1984, is Erikson dealt with in respect of linking his stages to those of homosexual identity unfolding. The reason is easily apparent: Erikson believes that homosexuality is part of a retarded or arrested form of sexual development, and that sexual growth culminates in the Utopian state of heterosexual marriage. An uncomfortable parallel exists in the separatist philosophy that has beleaguered homosexuals for decades. However, as is argued in this book, homosexual identity can more correctly be seen as part of the wider spectrum of human sexuality in general.

▶ Descriptions of homosexual identity abound with truisms concerning the genesis of same-sex attraction, its stages, and its developmental patterns (De Cecco, 1981; King, 1984; Minton and McDonald, 1984; Thompson et al., 1973). However, little attention is directed towards the internal needs and responses of gay people in respect of their identity priorities.

▶ The fast flow of academic insights, coupled with gay activism, has induced yet another double bind for gay persons, since the acceptance of homosexuality as non-pathological by some does not necessarily imply either automatic self-acceptance by homosexuals, or societal acceptance of them.

Hence, for most gay people there is a continual disparity between internal and external levels of acceptance, and the assimilation of feelings and experiences. Internal *implosions* are often lost at the expense of external *explosions* and represent a major contributing factor towards the on-going crisis of homosexual self-oppression.

▶ Homosexuals have felt the need to defend themselves against the onslaught of accusations made against them. This collective defence, or form of projection, has often been generalized towards society. Society's interpretation of this homosexual heterophobia has been to reduce it to the level of a nuisance factor, and to initiate counter-attacks against homosexual behaviour. This point is illustrated by a memorandum submitted to the Select Committee on Health and Social Services of the British House of Commons, in respect of AIDS. The author, a medical doctor, in a diatribe against homosexuality, warned the British government of its evils:

The most urgent step to be taken is to break the pervasive grip by homosexuals on the information and disinformation which has emanated for so long from the journals of science and medicine, and from much of the media. Once this has been done, other scientists, doctors and politicians can assess accurately the reality of the situation (Seale, 1986:11).

The subjectivity and distortion which sometimes surround homosexuality often also characterize the discussion of human sexuality generally. An un-emotive operational definition of human sexuality is that human sexual conduct is 'the expression of the physical and psychological experience of sexual desires and/or sexual usage, for physical and/or social ends' (Hart, 1979:6). Such a definition facilitates the objective study of human sexuality. Moreover, since sexuality does not exist in a vacuum, understanding can be further promoted by examining human sexuality in its many dynamic cultural relationships. For this reason, the concept of homosexual identity is discussed in this book within the systemic framework of interaction between society, the homosexual, and the homosexual sub-culture.

Homosexual identity

Homosexual identity and its development is most appropriately addressed within the broader parameters of human sexuality (Elphis, 1987; Rojek et al., 1988; Stein, 1988). It can be contended that historical enquiries, which have often viewed homosexuality in isolation from sexuality generally, have obscured the true understanding of homosexual identity. The first distortion is noticeable in definitions of homosexuality. The majority of texts, including common works of reference, define homosexuality parochially as a sexual proclivity between people of the same gender.

These definitions embody the generally held belief that homosexuals are creatures of sex, while heterosexuals are creatures of normal interpersonal relationships. The tacit message is that homosexuality is abnormal and without the purpose of a relationship i.e. that it consists of overt sexual behaviour only, without the relationship content that overt sexual acts normally imply in human society. Ironically, this opinion is widely held by *both* homosexuals and heterosexuals.

A second confusion lies in the association of homosexual behaviour with mental abnormality. While it is true that in isolated instances homosexual behaviour can be exhibited by persons with personality disorders, the same is true of heterosexual behaviour. The overwhelming majority of homosexuals are mentally healthy people.

The third obfuscation relates to identity. The literal meaning of identity is 'individuality' or 'personality'. Therefore it is logical to conceive of identity as an essence lodged within and determined by the individual.

Identity is not synonymous with gender (Hart,1979). It must be clearly stated that while gender differentiates the male and female physiological attributes that are genetically inherited, identity relates to gender only in so far as the male or female physiology is incorporated into the psycho-social structure of the individual. Hence, Sarrel and Sarrel write of identity not as 'a body reality, but the "perception" of reality — the personal meaning — for a given individual. The composite of these perceptions is called the "body image", and it is made up of unconscious, preconscious and conscious elements' (1979:23). The following definition of homosexuality embraces both gender and homosexual identity:

> Homosexuality is seen as a broad spectrum of psychological, emotional and sexual variables in a state of interplay between two [sic] persons of the same gender. Homosexuality is not only sex attraction between two [sic] persons of the same sex, but also includes: (i) an emotional as well as a physical bond, (ii) a fantasy system, and (iii) an element of symbolism, eroticism and sexuality. Homosexuality can be experienced in different degrees (Isaacs and Miller, 1985).

De Cecco and Shively complement this definition with features that include identity factors as well as sexual conduct within the context of persons in direct relationship with one another. They say that homosexuality may include 'beliefs about biological sex, femininity and masculinity, complementarity, exclusivity, sensuosity, intimacy, and permanency' (1984b:2).

Identity issues have been discussed by Erikson (1956, 1959, 1963, 1968), Freud (1977), Kinsey et al. (1948), Mead (1934), and Stoller (1969). In particular, the symbolic interaction approaches advocated by Plummer (1975) and Mead (1934) provide a paradigm for the understanding of identity. Identity is perceived as an evolution of experience proceeding through an increased awareness of a person's ability to harness the attitudes of others towards his/her own attitudes and behaviour. In other words, a dual form of differentiation occurs. The 'self' is seen because of competition with the 'other', and the owning of one's experiences is based on an external reality which ultimately becomes part of the internal being. This form of symbolic interactionism emphasizes that the negotiation of the *self-other* and the *I-me* dichotomies requires the structuring of experiences, mediated through the social process, and the subsequent (if not simultaneous) internal acknowledgement of the process. This point is extremely important, as it has direct relevance to a subsequent chapter of this book dealing with sub-cultural influences on homosexual identity development.

For the homosexual person, a *duality of experiences* exists. He gains two sets of behaviours (self versus self, and self versus others), firstly from his immediate environment (family, friends, school, and community) and secondly from his direct or indirect exposure to the homosexual sub-culture. The work of Erikson can expand this view. Erikson speaks of the ability of a person to situate himself within a social context. The inability to feel comfortable within any given social context can give rise to a sense of alienation, leading to the prerequisites for a state of crisis. Thus identity is both a process and a structure for foundation. For Erikson, a psycho-social identity is a continuous exchange, in which earlier phases of identification patterns, present forms of competencies, and future aspirations are interdependent. He focuses upon adolescence as a crucial period for identity stability, and believes that the adolescent discovers identity through 'achievement that has meaning in the culture' (1968:228). Moreover, each culture nurtures its members towards some form of identity consolidation which is both appropriate to the relevant social structure and idealistically congruent and manageable for the individual. These issues are crucial to the interpretation of homosexuality identity formation advanced subsequently in this book.

The unfolding of homosexual identity may be described as a tripartite process. Firstly, the individual has to deal with his (or her) biological inheritance and come to terms with the masculine (or feminine) gender. This is chiefly facilitated within the boundaries of the family system, and is part of the person's social identity development, based upon the codes of the impacting family and community structures. The second process is the person's internal dialogue with himself/herself. The person's identity fantasies and expectations become a private matter, but occur within the parameters of his/her social *milieu*, or micro-culture. The third process, and perhaps the most powerful, is that of sub-cultural identity, which manifests at a later stage of psycho-social development (usually late adolescence onward), and which is dependent upon the successful negotiation of the previous two phases for some form of cohesive integration to occur. While Erikson sees 'identity crisis' as a *rite de passage* within the framework of adolescence, he believes in the remedial support of peers and close friendship cliques, so that a progressive sense of belonging and being within an experimental context helps to negotiate an identity. As will be discussed later in this chapter, this period underpins the hidden crisis of the homosexual person, whose sense of identity has often unfolded (internally) during the pre-adolescent years. Thus, during the formative and highly evocative period of adolescence, the person who is dealing with his/her homosexual sensibilities has also to contend with his/her identity features within a tripartite process.

Erikson's eight stages of identity development[1] are relevant for an appreciation of the *longitudinal* process of identity formation, and constitute guidelines for the comprehensive understanding of identity patterns. Erikson warns that the unsuccessful negotiation of one stage (e.g. 'intimacy versus isolation') can lead to the formation of a full-blown crisis at a later stage, with the surfacing of residual trauma from the earlier period. Thus people who, due to earlier problems of development, find their sexual desires unbearably threatening, will experience severe identity confusion. Similarly, people who find that their sexual feelings are not within the context of so-called 'normal' expectations will set themselves apart, knowing that they are *different*. This sense of difference needs to be examined.

As this chapter goes on to reflect, feelings of a sense of 'difference' emerge within homosexual people during childhood, and are often sustained throughout the period of subsequent life transactions. The sense of difference is in fact an *idiosyncratic* experience within the context of the individual's private world, juxtaposed with his external ego realities. The concept 'different' can be featured in homosexual identity in many ways, for example:

- different because of sexual activities/desires;
- different because society has labelled them as such;
- different because of a stereotypical fancy, such as being 'artistic' or 'sensitive';
- different because 'difference' has become personified as a way of life; or
- different because the gay sub-culture has accepted this sense of difference as an identity construct.

Troiden, in a study of gay identity acquisition in 150 homosexual males, reflects that the 'difference' is in fact a sense of 'apartness from more conventional peers' (1979:363). This sense of difference can be manifested by general alienation, a feeling of gender inadequacy, as well as warmth and excitement in the presence of other males. Minton and McDonald (1984) contend that the difference expressed by homosexuals during their childhood and adolescent years is a feeling of separateness as well as isolation, which combine to give rise to a sense of sexual difference. Sexual identity is more than a biological concept that makes the anatomical differences between males and females the pivotal distinction. The mistake that has been made in the past, based upon bio-genetic theories, is that the male-female differences have actually been described as psychological properties of the individual. Notable are those distinctions governing feminine or masculine attitudes, male or female behaviour patterns, and male and female sexual conduct. These differences have usually been taken to have a socially determined status or interpretation.

A major dilemma in examining identity is that confusion exists as to whether an ego identity is synonymous with homosexual identity. Theorists, including Cass (1984), Dank (1971), Humphreys (1979), Richardson and Hart (1981), Weeks (1981), and Weinberg (1978), have argued for a distinction between behavioural and cognitive issues. They express the belief that sexual preference, sexual orientation, and sexual identity are separate issues, and support Erikson's contention that ego identity must be seen as separate from sexual identity. Cass (1984) succinctly places this in perspective when she reminds us that, once identity is established, it becomes an integral part of the person, hence the belief that homosexual identity is in fact a true or real self. This owning of one's identity is supported to a large extent by the gay sub-culture, which holds that a homosexual identity is only true when the person can subscribe to the word 'gay'. Because of the all-embracing spread of features which govern a homosexual identity, ranging from ego assimilation to minority identity issues, Cass has the following to say:

> The multidimensional continuum approach suggests that homosexual identity may vary on any number of dimensions. There are [sic] a myriad of meanings that individuals can include in their perception of themselves as a 'homosexual'. A sound theory of gay identity must be able to incorporate within its proposals the multifaceted nature of identity. What is the content of the different aspects of homosexual identity? What is the relative importance of each component in different life situations and for different individuals? Which personal and social factors are influential in changing identity components during identity acquisition? (1984:116–17).

To place Cass's rhetorical questions within the context of this book, homosexual identity development is seen as a vibrant and ongoing system which is nurtured within the following structures:

▶ the apparatus of fantasies and daydreams;

▶ gender attributes including the processing of animus-anima issues[2]; and

▶ sub-cultural influence.

Thus cerebral fantasy, external reality, and specific sub-cultural patterns are basic to homosexual identity. It is further suggested that two overriding and interconnected features lead to confusion within and without the homosexual world: by emphasizing gender attraction at the expense of emotional or other forms of ideological expression, homosexuality and its influence on individuals has been enshrined in an edifice of sexual 'acting out' behaviour. This may be attributed to the fact that homosexuals have been considered sexual deviants (not *human* deviants) and so the messages and symbols of sex indulgent behaviour have become part of an enmeshment philosophy. Understandably, homosexual 'procreation' takes place via specific ritual determined by the sub-culture. Thus, to perpetuate the sense of identity, the sub-culture, with its entrenched system of messages and meta-messages, allows for the metaphorical procreation of the 'species'. Therefore sexuality, a core participant in the totality of the identity struggle, has been deified within the homosexual mythos. Homosexual identity will continue to be popularly perceived in sexual terms until this umbilical cord, which attaches the homosexual to the womb of the sub-culture with its sexual protocols high on the priority list, is placed in perspective.

The writers cannot deny (nor would they wish to) the importance of understanding both psychodynamic and learning theory concepts pertaining to human development and personality actualization — with particular reference to homosexuality. The id, ego, super-ego, collective unconscious, as well as family dynamics are basic to homosexual reality (Isaacs, 1979a). Nor can the emphasis of more recent (and relevant) sociological theories of culture, role, systems, and communication be cast aside. With this acknowledgement, the following points serve as a preamble to the holistic model of homosexual identity development offered in this book:

▶ The 'cause' of homosexuality, as in the last analysis the cause of every human condition, is no more and no less than the successful procreation between a man and a woman, resulting in offspring.

▶ The infant, with his/her biological inheritance, usually responds to a set of male-female images (Murray, 1968).

▶ As the homosexual cerebral schemata (often of unknown and undetermined origin) begin to mature, so the human ego begins to develop.

▶ The human ego has a dual task. First it has to decipher and deal with internal responses or sets of feelings that have, in Jungian terms, an 'animus-anima' component. Second, it must respond to and deal with the external male and female systems which confront it, and between which there might be a conflict.

A proposed model of homosexual identity development

The model of homosexual identity development proposed by the writers is presented in six developmental stages, with the relevant, but *relatively arbitrary* age profiles accompanying each stage. The model is based upon an extensive analysis of the literature (Stricklin, 1984, has been particularly helpful), upon the writers' clinical practice experience, and upon data gathered from a survey of homosexual males that is reported upon in detail in Chapter 7 of this book.

Although this model is homosexual-specific, its stages of identity development could equally apply to human sexuality in general (i.e. heterosexuality). Fantasies, notions of difference, identity confusion, and idiosyncratic styles of sexual expression are not unique to homosexual development. However, in order to elucidate the specific structure and patterns of homosexual identity growth, the descriptions relate to a homosexual ethos only. Therefore, like most developmental theories, this model has a heuristic value (Coleman, 1987; Weinberg, 1984), and thus does not conform to a definable reality. The main features of the model are summarized in Table 1.1.

Table 1.1 The Stages of Homosexual Identity Development.

Stage	Description + age spread	Fantasies	Sexual behaviour	Reality base	Self-esteem/ congruency
ONE	IDENTITY DIFFUSION 0–9	primitive	spontaneous play	internal only	latent but undeveloped
TWO	IDENTITY CHALLENGE 10–15	evolving due to puberty	experimental, auto-erotic	internal-external dichotomy, diffuse boundaries	confused and non-manifest
THREE	IDENTITY EXPLORATION 16–19	ambivalent	testing out, bisexual phase	indirect or direct exposure to sub-culture	anxiety, turmoil, self-esteem jeopardized
FOUR	BEGINNING IDENTITY ACHIEVEMENT 19–25	homosexual focus, idealized	exploratory searching behaviour	participates in 'coming out' crisis	chaotic, fluctuating with socially narcissistic components
FIVE	IDENTITY COMMITMENT 19–65	homosexual object (s), homo-erotic bias	searching + co-habiting	participates in sub-culture	self-acceptance
SIX	IDENTITY CONSOLIDATION 19–65	ownership of fantasies	bonding, confirmed + secure	life-style synthesis	self-actualized

NOTE: 1. Stages 3 to 6 are not necessarily in age-linear progression.
2. Stages 5 and 6 are dependent upon levels of commitment to the identity as well as significant support from others.

Stage 1: Infancy and early childhood (birth to 9 years) — identity diffusion

There is universal agreement among child development theorists that the first four years of life are crucial for development (Barker, 1983; Biller, 1971; Bowlby, 1970; Erikson, 1963; Freud, 1977; Klein, 1959; Mahler, 1971, 1974; Winnicott, 1965, 1971).

Klein and her associates in the object-relations school stress the importance of the internalized object in determining the alteration of instincts, intra-psychic conflicts, and psychic structures of the infant. Klein states:

> If we look at our adult world from the viewpoint of its roots in infancy, we gain an insight into the way our mind, our habits and our views have been built up from the earliest infantile fantasies and emotions to the most complex and sophisticated adult manifestations (1959:302).

Mahler (1971) provides added insight into early bonding experiences between the infant and his/her caretakers, and postulates a separation-individuation process. This involves a complex and primitive psychological sequence that evolves during the first two years of the child's life. *Separation* occurs when the child moves away from a psychic fusion with the mother (or caretaker), and *individuation* represents the steps that lead to the development of an individual's personal and unique characteristics.

Wirz (1988) views the transition period as an uncompleted process, but emphasizes that the favourable negotiation of the separation-individuation sequence is thought to lead to *psychological birth*. She suggests that this negotiation helps to facilitate the development of adaptive capacities, the acquisition of identity, and the resources for mutuality in human relationships. Furthermore, she stresses that the qualities of trust, compassion, and congruency are acquired through significant exchanges between the child and the family (or significant others).

Thus, the infant responds primitively to the cues offered to him by his caretakers, and this sets the pace for the beginnings of the process of his individuation or, in object-relationship terms, the beginning of a sense of self perceived as separate from others. In studies determining the onset of homosexual behaviour (Bieber, 1962; Freund and Blanchard, 1983; Stoller, 1969), a common feature has been an exploration of the labyrinth of responses from homosexuals to male and female parent objects, resulting in a range of hypotheses. Among these have been theories of bisexuality, castration anxieties, anal fixations, and inversion (Murphy, 1984). There is no doubt that the transactional experiences between a homosexual person and his caretakers have predisposing elements towards later patterns of sexual identity, sex role preference, and sex role orientation, although a distinction must be made between early feelings of same sex attraction and feelings of homosexual sensations. The former are usually expressed at around puberty, a fact corroborated by the survey reported on in Chapter 7. The latter are usually expressed (on recall) at approximately four to five years of age.

This circumstance can be explained within an existential framework, and leads to the special notion of *homosexual idiosyncracy* in respect of the ontology of homosexual development. There is no semantic or emotional interpretation by the child during this early phase. Recall descriptions given by survey respondents and clients in therapy range from 'I felt different ... I recall a bubbling sensation in my stomach when in the company of boys and men', to 'I knew I was special, and felt warm towards men ... especially my father'. This period, too, has been associated with extraordinary behaviour, such as cross-dressing, playing with opposite sex toys/objects, and acting out behaviour, usually embodying fantasy and creative outbursts. These behaviours can be an outlet for the child in 'telling' his caretakers or significant others of his special needs. This stage is usually accompanied by a primitive and recurring fantasy, as exemplified by a client who recalled: 'At the age of six, I remember boys dressed in black and white suits, similar to penguins. I remembered that they were boys!'

Thus fantasy, which includes experimentation with clothes, objects, books, and intimate contact with others, is the first spontaneous clue to the onset of sexuality. This phase was described by one respondent to the survey as 'exciting, cathartic and frustrating'. 'Frustrating' is the key word here, for it could be extended into the first experience of fear. Fear (or intimidation) can be a direct response to adult or peer group disapproval or ostracism. It is not the child who cannot cope with his sets of behaviours, but rather his adult and peer group world who perceive the behaviours to be out of the ordinary. Thus the child obtains feedback from an external source, not having access to his own sense of morality for it has not yet developed. He is either punished — or, paradoxically, rewarded — for his outward manifestations of this idiosyncratic behaviour.

A major feature of this stage is that, although outward behaviours are manifest, the fantasy process usually belongs to the child. When he is reprimanded or scolded for untoward behaviour, his sense of inner worth and early experiences of unfolding identity are bruised. The child usually remembers this hurt as being objectionable, and the initial responses to negative identity feelings are stored. Furthermore, a thin membrane distinguishes sexual identity development from sexual confusion. As an example, attention is drawn to the fact that certain cultures, as well as certain traditions, promote gender behaviours contrary to the prevailing norms. Examples include masculine names given to females, or parents desiring a daughter, and dressing their son in girl's clothing. Such parental action may or may not have a bearing on the child's ultimate sexuality, since the fantasy does not come from within the child's internal framework, but is imposed externally. In this respect, some studies have argued that cross-dressing and sexual confusion do not necessarily predispose towards sexual identity (Ross, 1983c; Sipova and Brzek, 1983).

Stage 2: Puberty and early adolescence (approximately 10 to 15 years) — identity challenge

During this stage, which is marked by the onset of puberty (and, according to Erikson (1968), is vital for ego integrity), the enquiry process develops, whereby semantic, cognitive, and behavioural components are grafted into the fantasy repertoire. The fantasy, which incorporates antecedent images and experiences, is now often accompanied by accidental or deliberate masturbation and other forms of sensate focus arousal. The fantasy and its agenda of images is more evolved, with a direction towards same-sex objects or erotica, depending on the level of sophistication of the adolescent, as well as his exposure to a 'sexual object universe'. The beginnings of a sense of morality emerge (a super-ego construct) including the belief that his internal and private world is wrong. Of note here is the adult world's attitude towards masturbation, as well as 'wet dreams'. If the adolescent is chastised about masturbation, this can only compound his sense of shame, wrongfulness, and guilt. The ego's exposure to codes of sexual ethics, socio-biological preparations for boy-girl role behaviour, and adult heterosexual models, is *out of alignment*.

One client, for example, who from the age of eight years lived with his father and a male lover, was expelled from school when, during a 'sex guidance' lesson, he asked the teacher what was wrong with boys holding hands with boys, when she was referring to courtship patterns between adolescents. The internal reality does not correspond to the adolescent's external experiences (or messages). With little or no chance of his internal dialogue being validated by others, access to his fantasies and/or special behaviour from the past, as well as the thoughts that accompany his autoerotic arousal, give rise to a sense of confusion. The opportunity to share these experiences — unless within the framework of so-called adolescent sex experimentation with a significant person who can verify his fantasies as being non-destructive — is minimal. Periods of internalization occur, corresponding to the feelings of 'separateness' mentioned earlier. This sense of separateness is often misdiagnosed by clinicians as anti-social, or conduct disorder behaviour. The significance of this period is frequently dismissed or underestimated by society, in the belief that adolescents who indulge in same-sex behaviour are simply 'going through a phase' (Colgan, 1987).

Because of the prominence placed on adolescent behaviour, and the added stress of a twilight existence bridging childhood and adulthood (so brilliantly described in William Golding's *Lord of the Flies*), adolescents have to cope with demands made on them by their immediate culture as well as their own sexual itinerary. A key feature during this period is an acknowledgement of sexual attraction. Within the homosexual ethos, this attraction is expressed through male imagery, usually fixated upon the erotic zones, such as the crotch, thighs, buttocks, and chest areas.

One client recalled:

When I was 15, I used to watch a grown-up man in the apartment block opposite. He would come home late at night, and I would stealthily creep onto our balcony to watch him undress in front of his window. My parents were asleep, and I did this for at least six months. I would go back to bed and masturbate. Later I felt that this was so wrong that I tried to punish myself each time I found myself looking at men. I can recall taking mustard mixed with salt water to make myself sick.

The imagination does not have to be extended to understand the sense of anguish experienced by the person who recounted the above vignette. Of importance is the dread of being discovered, and hence the recurring fear of punishment or abandonment. Most clients interviewed by the writers endorsed this fear, in particular when being exposed to men in public, such as at swimming pools, showers, and urinals. Some described an inability to urinate in front of other men, and how they feared that other men would see that they were 'different'. In particular, they experienced a basic fear of uncontrolled erection, thus giving their secret away. Adolescence is a period of exploratory behaviour, including a sense of sexual urgency and excitement. Many homosexuals (and this can be confirmed by major studies, as well as the writers' own study reported upon in Chapter 7) can pinpoint this stage as the conscious genesis of their homosexual development.

Stage 3: Middle to late adolescence (approximately 16 to 19 years) — identity exploration

This period[3] incorporates active testing out behaviour, primarily within the context of bisexual exploration. A sense of doubt permeates this stage, with the adolescent vacillating between his inner moral interpretations and his basic sexual legacy. The majority of people on the writers' case-loads reported, in their psycho-social histories, having varied sexual contact with girls/women as well as boys/men. Both the data of the writers' survey and the literature in general suggest that the age of 18 years is significant in the adolescent's sexual life. Descriptions of a meaningful encounter with a person of the same sex are often given. A few clients recalled traumatic incidents, including rape, seduction, and poor sexual responses such as secondary impotency, premature ejaculation, and 'inferior' penis size.

> One client, aged 22, who was raped in a university residence when he was 18, 'blocked' his sexual responses towards men. He feels cerebrally homosexual, but claims to experience 'frigid' behaviour when confronted by gay men. Upon exploration of his fantasy repertoire, he was able to testify that it included two persons having sexual intercourse: one male, one female. Upon further probing, the female turned out to be the client *in disguise*. His distorted self-image, coupled with his desire for anal penetration, represented passivity and lack of control. This did not correspond with his image of homosexual men, who ought always to be macho ('butch') and strong.

> Thus this person perceives himself to be neither male nor female, and not homosexual either. His sense of 'ideological' correctness has not yet embraced so-called androgyny. He is thus living a *non-sexual* existence coupled with intense confusion, akin to phantom homosexual behaviour or pseudo-homosexuality, as expounded by Bieber (1972) and Ovesey (1955, 1969). This state has jeopardized the person's acknowledgement of his feelings, and has led to the beginning

of homosexual as well as heterosexual phobias.

The stage of Identity Exploration, as reflected by traditional (orthodox) descriptions of adolescent sexuality, is governed by adolescent turmoil, rebellion, further physiological changes, and new patterns of decision making. Competition, narcissistic features, and masculine assertion form part of it. As the individual gains intellectual and emotional maturity, the fantasy, coupled with greater detail of sexuality in general, becomes specific and less free-floating. The person is in control of his daydreams and fantasies. If they correspond with the content of dreams accompanying nocturnal emissions and other dreams which reflect sexual imagery, then they serve to confirm his identity. This period, if not reconciled, can reach a traumatic pitch, and a period of stopping masturbation and other self-stimulation is known to exist. This so-called 'temporary moratorium' is common, and a direction towards opposite-gender behaviour takes place, including dating and the pursuit of overtly masculine activities, including sports such as rugby. The psychological defence of reaction formation ensues. Within the gay vernacular, this process is commonly referred to as 'being in the closet' and represents an interim denial of homosexual identity.

Testing out behaviour occurs with peripheral exposure to the gay scene (such as the reading of certain books, and exposure to media containing homosexual content), or privately within the confines of daydreams. It is not uncommon for adolescents to search actively for confirmation of their identity, and a number of clients have reported that they hitch-hiked with the purpose of receiving lifts from strange men, so as to land up in a compromised situation.

This stage is governed by one notable feature: *confusion*. The confusion is not about identity *per se*, but rather how to express identity needs and wants within a largely hostile external environment. Furthermore, the fantasies and/or experiences are flooded with sexual images (such as figures of hero-worship), which in turn are often confused with sexual desire. At this stage, a strong need is experienced to formulate a sexual sense of priorities.

Stage 4: Late adolescence to early adulthood (approximately 19 to 25 years) — beginning identity achievement

With adult independence in sight, with job opportunities, and with the lessening of family ties, the person is able to embark upon some form of fantasy consolidation. Interaction with homosexual objects becomes far more intent, and strategies are sought to experience or express such interaction. Fantasies, irrespective of masturbation, take on greater impetus, and the sexual object of desire is expanded into a Gestalt composite of man or men (Isaacs, 1979a).[4] The searching for inner reality and personal truth becomes formalized and, depending on geographical and social circumstances, accidental or planned interaction with the gay sub-culture takes place. It must be remembered that in South Africa, access to homosexual social institutions is usually restricted by law to adults. Such institutions also impose their own strict protocols for admission, in order to avoid clashing with the authorities. Thus fulfilment of expectations usually takes place in private, via the clique system, or in well-known meeting places (for example, the beachfront in Cape Town).

From a psychological point of view, this stage often determines the person's projection of an ideal fantasy image, usually referred to as a 'type' in clinical parlance. Types are part of the person's search for ultimate identity, and are based both on socially determined images and on individual psychological needs, such as the need for a 'father-figure'. Throughout the duration of sexual interaction with self or others, persons will reflect their desires, not only of body types or stereotypical images, but also according to their levels of narcissistic development (Friend, 1987; Gonsiorek, 1982), their previous bonding experiences with significant others, and their need to test out the variables of power (Foucault, 1976; Hearn and Parkin, 1987; Silverberg, 1985), intimacy, and control.

In this respect, intimacy serves to affirm interaction between individuals, and incorporates levels of trust, with particular reference to self-disclosure, faith in dependability, and affirmation of self-esteem.

This stage is usually marked by *homosexual identity confusion* during 'coming out'. To illustrate this, and to demonstrate the essential features of fantasy — power, intimacy, and control — Figure 1.1 depicts the masturbatory-fantasy sequence as it relates to homosexual identity development during the 'coming out' process.

Unless the penis is erect due to spontaneous but non-sexual arousal such as morning or nocturnal erections, the person needs to experience tactile, auditory, or visual stimuli in order to engage the autonomous nervous system response which facilitates the flow of blood to the penis. During the period of homosexual identity confusion, the psychic trauma resulting from the recognition of homoerotic stimuli can precipitate secondary impotency, ejaculatory problems such as premature, delayed, or non-orgasmic discharge,

FANTASY CONTENT

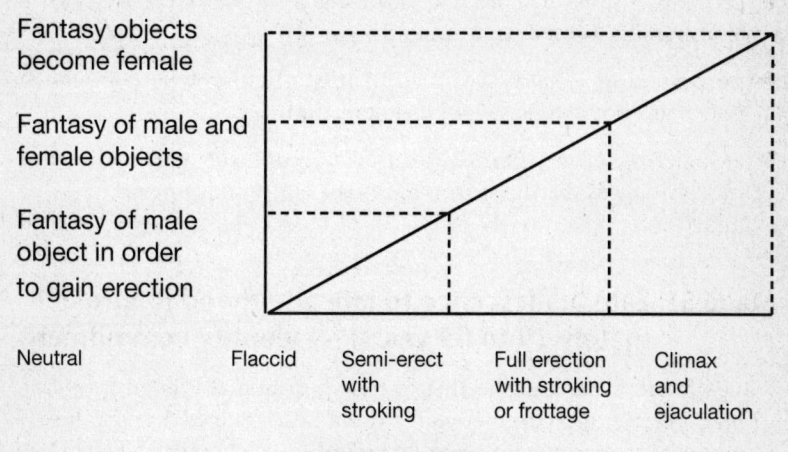

Fantasy objects become female

Fantasy of male and female objects

Fantasy of male object in order to gain erection

Neutral   Flaccid   Semi-erect with stroking   Full erection with stroking or frottage   Climax and ejaculation

PENILE STATE

Figure 1.1 The homosexual fantasy-masturbatory cycle during 'coming out'.

or prolonged ejaculatory delay causing penile trauma (such as abrasions or bleeding), boredom, fatigue, and anxiety. Thus, in order to gain an erection, the person has to fantasize upon a desired object. Once erection has been achieved, one or both of the following scenarios can ensue:

▶ The desired homoerotic image is replaced by a female object. During this process, a vacillation may occur whereby male images may intrude upon the now primarily female object. Furthermore, the male object could be the person himself or a voyeuristic impression of himself observing the male-female act from a distance, thus maintaining a *safe* homoerotic fantasy.

▶ At the height of the masturbatory act, when orgasmic sensation is near its peak, the *female object* is consciously fixated upon. During and after orgasm, the person associates this experience with heterosexuality. The initial stage of getting an erection by using male imagery becomes minimized or even temporarily repressed. Consequently, a form of self-regulated behaviour modification is incorporated into the person's internal fantasy system.

The above process can complicate the development of a homosexual identity by:

▶ causing a masturbatory moratorium;

▶ promoting guilt and the fear of losing self-control by the person abandoning himself to homoeroticism;

▶ perpetuating identity confusion, in particular where intimacy issues are not actualized with others; and

▶ associating *homosexuality* and *homoeroticism* with feelings of *powerlessness* in the form of weak, effeminate, and passive images.

Stage 5: Late adolescence to late adulthood (approximately 19 to 65 years) — identity commitment

Stages 5 and 6 expand the first four stages, hence it is not possible to give precise age parameters for them; also, Stages 5 and 6 incorporate most of the determining factors of the previous stages, therefore the stages should not be seen as mutually exclusive. There is overlap between them, and it is thus possible for an individual to fluctuate between one stage and another.

Stage 5 is regulated by a process whereby the person learns about, or determines what conditions have permitted him to acknowledge '*I am homosexual*'. Dank (1971) believes that this period includes dealing with certain issues. First, the social context of 'coming out', which Hooker (1965) describes as the 'gay debut', is examined. Next, cognitive changes are reflected upon. There is an examination of the implications of identity and self-acceptance, which Gagnon and Simon (1974) refer to as the individual's self-recognition of his identity as a homosexual within the context of exposure to the homosexual network. There is also a response to public and self-labelling. Moreover, thought is given to dealing with the 'closet' syndrome,[5] in combination with the role of acquired knowledge specific to homosexual culture.

This stage captures the psychological prerequisites that determine a basic identity within the fabric of the gay sub-culture. Active searching for role models, as well as consolidating sexual expressions of intimacy, form part of the testing out period. This phase, too, is the precursor to episodes and problems which may manifest in crisis proportions, often referred to as 'the coming out crisis'. A period, or several periods, of *identity intolerance* is experienced. Webster (1977) maintains that this may lead to a full-blown identity crisis. The nature of this crisis is usually determined by a deliberate attempt to disinherit homosexual patterns because of internal conflict and a fear of the sub-culture.

Stage 6: Late adolescence to late adulthood (approximately 19 to 65 years) — identity consolidation

This period reflects the consolidation of the diversity of experiences within homosexual behaviour. Attempts are made to 'couple', hence romantic relationships and a sharing of domestic alliances feature at this stage (McWhirter and Mattison, 1984). Conditions of love and emotional and sexual commitment are necessary for this stage to evolve fully. Homosexual identity is usually comfortable within this period, but this does not necessarily imply total acceptance of the self, nor of the homosexual persona. Internal re-evaluation of personal codes of behaviour usually takes place, in direct response to external factors. These external factors (which impinge upon growth and which determine the person's ability to evolve from a field-dependent to a field-independent person[6]) are examined below in terms of the parameters of sub-culture, sexuality, family system, relationships, AIDS, and political issues.

Sub-culture

The sub-culture has a deterministic quality for homosexual identity. Individuals who interact with the sub-culture frequently claim to be satiated by its socially and sexually incestuous activities. This is an apparent form of inverse victimization, since the sub-culture does not have the sustaining features to maintain homosexual homeostasis.

The sub-culture is composed of ritual, or rite. According to Turner, ritual is marked by three distinct phases: separation, margin, and aggregation. During the intervening period of sub-cultural acquisition, the characteristics of 'the ritual subject (the passenger) are ambiguous; he passes through a cultural realm that has few or none of the attributes of the past or coming state' (1969:80).

Traditionally, the sub-culture has institutionalized a diversity of homosexual behaviours, many of which substantiate or reinforce a state of identity — or promote a sense of identity collapse. Some of these behaviours include sexual mores, language, dress, formalized meeting places, literature, and a sense of political ideology.

In fact, the sub-culture is an osmotic force that permeates the enquiry profile of the gay 'passenger'. Two levels of enquiry exist.

The first takes the form of direct interaction with the institutions of the sub-culture. The second paces the individual who has not come to terms with aspects of his identity in accord with his own needs i.e. it concerns the 'closet' person or the coming out syndrome in action.

Kenneth Read (1980), in *Other Voices*, succinctly describes the gay sub-culture in respect of its ambivalent profile for the gay person. He states:

> The gay community and gay culture are patent misnomers when applied to the population at large. There are some gay communities and there is a minimal lore which is understood by a large number of homosexuals. There are, however, many more specialized lores that are not shared, and that are mutually exclusive. Paradoxically too, the constant misuse of the terms 'culture' and 'community' may be a disservice in the long run to the achievement of laudable goals, fostering the long-standing myth that homosexuals generally are members of a subversive conspiracy (1980:8).

Thus, as will be shown subsequently in Chapter 3, the sub-culture has implications for identity growth, in that it both fosters self-actualization, and perpetuates self-oppression.

Sexuality
Gender identity is more than simply 'sex appropriate' behaviour or secure sex role identity. It is more than just an image of the self as 'masculine' or 'feminine', and an acceptance or understanding about what membership in that group involves. It is also about the integration of a person's sexual impulses and urges into this role. It must therefore necessarily include a person's self-awareness and self-acceptance of sexual orientation, and an ability to express this meaningfully with others (Isaacs, 1979a:13).

Sexuality, therefore, is behaviour. This behaviour, so much a part of the human identity, is based upon gender identity constructs (Stoller, 1969). Mannerisms, attitudes, and attributes are acquired from infancy into adulthood, and are stored in the sub- and collective unconscious. This constitutes the individual's collective experience and, although its baseline is formed in early development, it is nevertheless open to change. These concepts have been confirmed by Malyon (1982b), who contends that homosexual self-labelling, together with simultaneous awareness of homo-erotic desire, occurs during the earlier stage of adolescence. Because of social disapprobation, conflict arises in parallel with the suppression of homosexual fantasies, only to surface later — even decades later — in the form of unfinished tasks. These can cause confusion in respect of the four separate but interrelated variables that comprise gender identity, namely biological gender, sexual identity, social sex role, and sexual orientation.

Gender identity is usually verified through the medium of ongoing sexual thought (internal dialogue) with accompanying behaviour expression. If behaviour outlets are denied, forbidden, or 'taken away', the individual might regress. But the conflict might resurface at any moment during his life, with a distorted and painful crisis identity. Gershman (1983), in a sensitive critique of coming out published in the *Journal of Psychoanalysis*, reflects on this point and relates the stress of coming out to improper completion of the process of individuation during childhood. His use of the word 'improper' can be extended to the social context of behavioural controls upon the child, the consequences of which can emerge in later years as sexual despair coupled with conflict. A case vignette from clinical practice illustrates this point:

> A 35-year-old man of Southern European origin had been caught masturbating by his nurse when he was seven years old. After her severe chastisement of him, she pointed out the ruined shell of a nearby burnt out church. She instilled in him the wrath of the devil, and, as the client recalled, said 'If you continue to do that naughty thing, you too will be burnt by the devil and be forever locked in the flames of hell'. The verbal flagellation by the nurse, who represented an object of love for the client, induced severe guilt about masturbation. Early adult sexual experiences with men precluded mutual masturbation, and when masturbation was the only form of sexual expression, the client experienced secondary impotency. His form of sexual comfort was through oral or anal sex only. With the advent of HIV infection and AIDS, his recourse to these forms of sexual expression was restricted, since masturbation was advocated as a 'safe' (or safer) means of sexual activity. He sought help, because his present partner insisted upon mutual masturbation, causing his feelings of guilt to recur, together with impotence.

Family system
As Erikson says, roles 'grow out of the third principle of organization, the *social*. The human being at all times ... is organized into groupings of geographical and historical coherence: family, class,

community, nation' (1963:36). Within the gay collective, a disparity between self-acceptance and external acceptance is found, and many homosexual people choose to live at some geographical distance from immediate family in order to avoid conflict over their life style. This lack of family support, caused partly by the reluctance of gay persons to discuss (but not necessarily to disclose fully) their identity issues, can preclude a sense of understanding. Moreover, if some other person or persons, not of the family of origin, provide unqualified support to the homosexual individual, this may further inhibit him from sharing his joys, discoveries, and pains with those who ostensibly ought to be even closer to him (Cramer and Roach, 1988; Hammersmith, 1987). This situation may account for the symbolic family network that many gay persons create for themselves within their immediate sub-cultural interactions. The feature of symbolic or 'alternative' family structure is discussed more fully in Chapter 7.

Lack of involvement with family of origin can distort the gay person's sense of reality, induce long-standing anger and resentment, and negate a form of free expression — a component necessary for congruent behaviour. This can be illustrated from the following client situation:

> A client in his late twenties entered therapy in order to resolve a five-year-old relationship dispute. Part of the historical enquiry revealed a minimal external support system. His parents, who knew he was gay but who never commented on his life style were, according to him, unavailable for support. During the entire period of his cohabitation with his lover, his parents were never invited to the house. Similarly, his lover never met his parents or interacted with them in their home. He believed that his parents would 'freak out' if they saw two men living together, sharing a bed, and living like 'straight' people. The client, after discussion, acknowledged that keeping his parents away not only reinforced the estranged behaviour, but fed their fantasies that homosexuality is uncomfortable, distasteful, and secretive. This issue linked dynamically with the client's own sense of personal resolve. Therapy highlighted his sense of personal discomfort, and his inability to acknowledge homosexual issues which had a direct bearing on his relationship.

Relationships

There is no doubt that homosexual bonding within a relationship context plays an important role in gay identity. The relationship has both growth and saboteur features (McWhirter and Mattison, 1984). Although homosexual relationships have been described longitudinally, with a chronological sequence attached to the various stages, such as blending, nesting, maintaining, building, releasing, and renewing (Ibid., 16–17), special attention needs to be given to the sense of *urgency* in which homosexual relationships, particularly within the sub-culture, are negotiated. There is a strong desire to emulate heterosexual behaviour: in fact, this is one aspect that has been stereotypically described for years — the passive male coupling with the active male (George and Behrendt, 1987). With the gradual dispersion of myth, an extensive network and variety of relationships have been revealed, ranging from traditional dyads, to open relationships, to bisexual three-somes, to community or multiple relationships, as well as those which may simply be described as libertine. However, the point is that homosexual relationships, however perceived or experienced, have the task of *validating* a human existence. Such relationships are laboratories for the testing of sexual preferences, a sense of the erotic, fantasy consolidation and/or experimentation, as well as the emotional-spiritual capture of love and commitment. The urgency of couple-bonding is reflected in the need to prove a sense of completeness, in order to ameliorate the message of internal repression so often imbibed by homosexual persons. Rapid relationship negotiation, however, often results in equally swift and often terrifying dissolution of the relationship, followed by a rebound or reattachment episode. Hence, unless developmental stages prior to entering the relationship have been successfully negotiated, the relationship might prove to be the testing ground for a collection of 'unfinished business'.

Relationships and the sub-culture are inextricably intertwined, thus impacting upon identity. The sub-culture provides a powerful direction for the homosexual actor. Its ethos paradoxically both supports and undermines relationships within the context of sexual attraction, youthful conquest, and competition. The following extract from a clinical record illustrates the sub-culture's more negative impact:

> A 40-year-old man who had terminated a three-year relationship with an 18-year-old youth sought solace from a gay friend. The friend's immediate response was that gay relationships don't work, and that what he (the client) needed was 'a stiff drink at the bar and a good fuck' to get over the loss of his friend. The crisis, as introduced to one of the writers by the client, was not

the loss of the relationship *per se*, but the attitude of the friend, which symbolically represented the code of conduct of the gay scene. The client's sense of integrity, his sense of identity, his valid expression of grief, and mourning the loss were disallowed.

HIV infection and AIDS
There is no doubt that HIV infection, as well as AIDS, are contemporary influencing factors on the growth of homosexual identity (Isaacs, 1987a). Homosexuals (and others) are being forced to re-examine their priorities, relationships, and modes of sexual expression. In addition, AIDS has induced a scare or panic syndrome (Isaacs and Miller, 1985) in those who need to confirm or explore their identity through the medium of sexual intimacy. New defence mechanisms, often expressed at a conscious level, are being used by young and old alike. As discussed in detail in Chapter 5, this is a nightmare for therapeutic clinicians, for AIDS has blocked the avenue for talking about and dealing with intimacy (Harowski, 1987). Because a strong element of retribution and societal blame is re-emerging, issues of sexual identity and homosexual concerns that were previously taken for granted are now being unleashed in the clinician's consulting rooms, as well as in everyday interaction between people.

Political issues
In South Africa, as in some other parts of the world such as the Eastern Bloc, homosexuals feel a schism in their identity in that they have no positive legal recognition. Cumulative episodes of pain are reflected in their dialogues when they compare themselves with heterosexuals, whom they believe to be 'public' persons as opposed to homosexuals, who are described as 'private' or clandestine persons. The absence of an affirmative legal identity is worsened by a sense of self-oppression, a fear of blackmail, and a fear of public exposure leading to rejection, or, even worse, criminal action against them. The issue is further compounded by a strong contemporary movement to politicize gay issues. Increasingly, gay people are expected to identify with the 'social activists' within their ranks, and this has ramifications for identity. Those who fear public exposure at any level other than within the safe confines of the homosexual network disapprove of any form of gay activism because of its revealing intent. Moreover, there is a clash of political ideologies. There are also those people who invest energy in political activism at the expense of their own emotional and sexual

determinism. An example is sometimes found in 'political' lesbians, who are feminist and often Marxist in their outward philosophies, and who adopt a lesbian profile not because they necessarily possess the ontology of lesbianism from an early direction, but as a political statement directed towards men and the male concept of power and control. It is, of course, difficult to generalize within this particular area, but a similar division sometimes exists among gay male activists, who use activism as a form of surrogate sex, and avoid dealing with their own personal sexual needs. However, one also comes across gay and lesbian activists who have resolved their basic identity struggle, and who therefore have the emotional capacity to invest their energies in efforts towards meaningful change.

The six-stage model of homosexual identity development which has been described must be interpreted with elasticity and flexibility. Each stage can be viewed as a separate entity, but the connections and overlaps between them must be recognized. Furthermore, the first three stages are seen to be prerequisites for the satisfactory negotiation of Stages 4 to 6. It is not uncommon for a person who is theoretically placed in Stage 3 (for instance, a 16-year-old) to be able to consolidate aspects from the range of behaviours described for each stage. Thus, heed must be taken of Cass's warning: 'Homosexual identity must be seen to be *fluid*. An individual is seen to be in a state of continuous being. Therefore, identity can never be "what is", only "what is becoming".' (1984:120) She also challenges the 'essence' theorists, such as Berger and Luckman (1966), De Cecco (1981), and Plummer (1981a), who have on occasion described homosexuality within a fixed model, failing to recognize the process and diversity of identity.

Homosexual-linked behaviour patterns

The model offered in this chapter is a means of conceptualizing the development of homosexual identity. However, there are also behaviours which are not always homosexual, or not homosexual at all, but which are nevertheless *associated with homosexuality*. These homosexual-linked behaviours require identification and acknowledgement within the parameters of identity development. They include transient, situational, and accidental homosexuality;

bisexuality; transsexuality; transvestism; androgyny; and heterosexual homosexuality.

Transient homosexuality

By 'transient' homosexuality is meant a short period of experimentation, during which the homosexual transaction does not fulfil the emotional and social expectations of the person. Transient homosexuality is usually experienced by some people as part of a general inquisitive enquiry into sexual behaviour, as well as by persons who believe that it is avant-garde to be able to admit to a homosexual experience, often within a fringe or alternative cultural context.

Transient homosexuality is also experienced by persons who are confused overall about their sense of identity, and who may use the conscious and deliberate choice of experimenting with homosexuality in an attempt to advance their identity.

'Homosexual panic' as described by Bieber (1972) and Ovesey (1969), as well as by authors investigating relationships between mental disturbance and homosexuality (Frosch, 1981; Lester, 1975; Miller, 1978), is a reflection of the nature of transience within the context of pseudo-homosexuality, excessive anxiety, and ego-dystonic disturbance. Miller describes persons diagnosed as mentally ill (e.g. schizophrenics) as rarely being openly homosexual; their sexual identity conflicts are more typically expressed as homosexual fears, often projected as accusatory auditory hallucinations. He concludes his study by verifying that this type of transitory homosexuality, although repetitive in the context of the nature or the recurrence of the mental illness, is a form of hysterical behaviour as opposed to a longer-term psychotic process (Miller, 1978:113).

Situational homosexuality

'Situational' homosexuality is a conventional description, usually applying to a group (or population) of persons who are removed from society, and whose access to opposite-gender companionship is denied. Common examples are prison populations, ships at sea, men under arms, and (in the South African context) the Draconian system of migrant labour, which entails forced compound living, where thousands of men are separated from wives, other women, and families (Moodie, 1988). In such cases, the sex drive overrides the fantasy system, and a gay ethos is not apparent. Situational homosexuality, with few exceptions, expresses itself from without the gay sub-culture. Verbal reports from clients who were engaged in homosexual behaviour while incarcerated have revealed sets of heterosexual imagery, such as photographs, private fantasies, and verbal pornography (language of evocative sex) while engaging in anal or oral homosexual activities.

Situational homosexuality has a direct link with accidental homosexuality in certain circumstances. In Cape Town during and after a series of boycotts in 'coloured' and African schools,[7] a perceptible increase occurred in male sex workers. Although there was an even spread of 'coloured' and white men, some African youths were also involved in commercial sex work.

One of the writers conducted interviews with these sex workers ('rent boys') and their clients, all of whom voluntarily agreed to participate in discussions. Of the 10 sex workers, six were 'coloured', two were white, and two were African. The discussions revealed the following:

▶ All the 'rent boys' denied that they were gay.

▶ All had difficulty in actively engaging in a sexual act, and, as they were basically impotent with their clients, they usually performed fellatio or masturbated the client.

▶ They accepted being fondled or caressed by their clients, and on some occasions a spontaneous erection occurred (specifically if they referred to a female image). Some allowed a client, for an increased fee, to indulge in anal sex with them.

▶ All reported feeling comfortable with the opposite sex, and the whites said that they had current relationships with girl friends.

▶ Although all of the men were 'street wise', their major objective was to obtain as much money as possible. The fees that they charged ranged from R10 to R80, depending on the services provided.

▶ The two Africans, both under 16 years of age, were willing to engage in any form of sexual activity with their clients. Their experiences had soon rendered them familiar with some gay argot, behaviours, and ways of seduction, but their fantasies were in effect neutral.

Accidental homosexuality

Accidental homosexuality may be experienced in different forms and in different degrees. Usually it is manifest under duress such as rape or coercion, for example in a prison cell, or via group pressure. Toxic abuse may lead to uninhibited sexual drives, and episodes of same-sex sexual activities are known to have occurred under states of inebriation. The seduction by sexually more experienced men of others usually takes place under conditions of excessive toxic mismanagement. Again, accidental homosexuality may be seen as linked to situational and/or transient homosexuality.

Whereas accidental homosexual encounters may or may not lead to full-blown homosexual behaviour, in many instances the behaviour burns out if the experience loses its impact and meaning. In the majority of instances, the respondents to the survey reported upon in Chapter 7 recounted an element of discomfort with the onset of their homosexuality. A few interpreted 'accidental' events, such as molestation or seduction by older men, as a 'cause' of their homosexuality, rather than as a circumstantial event which gave rise to feelings within their idiosyncratic history of homosexual feelings and fantasies.

Childhood/adolescent homosexuality

Some form of same-gender experimentation often occurs in boys and adolescents. For some experimenters, this is a natural way of dealing with body space, of comparing overt sexual-gender apparatus, of responding to spontaneous arousal states, and of searching for male-female sets of differences. Adolescence introduces a frivolous sense of competition, and it is common for adolescents to masturbate mutually, or even 'bum rush' (anal frottage or penetration) as a form of enquiry, even suggesting dominance and power conflicts. For other adolescents, who have developed a homosexual sensation during infancy, childhood, and early adolescence, this period presents a different kind of experimentation. It is fraught with anxiety because their fantasies usually correspond to the homosexual onslaught.

Bisexuality

Bisexual behaviour is sometimes presented as being a period of transient exploration into *either* homosexual *or* heterosexual behaviour. For some it reflects an uncertainty in sexual orientation, so that same- and opposite-gender behaviours are both negotiated. A telling point in determining 'true' bisexual behaviour once more involves the exploration of the fantasies that accompany sexual activities. Some clients have reported more anxiety when relating to women, and have had the need to fantasize, or to use pornography (which includes males) to enhance their performance with women. Bisexual behaviour may also be reflected as a cult or fringe activity. With pop stars such as Bowan, Bowie, Boy George, and Elton John acknowledging bisexual activities, many persons have been released from their prisons of doubt.

Bisexuality may also be an ongoing form of sexual expression, and can be manifested by married couples (Bozett, 1981). A client of one of the writers believed that his marriage was solidified by his bisexual activities, and young men were brought home to be enjoyed by both his wife and himself. Finally, bisexuality may be expressed through sexual activity both in 'unisex' fashion, and symbolically. This issue has implications for the gay sub-culture due to the intrusion of the fringe culture, and it is taken up again presently in Chapter 3.

Transsexuality

The phenomenon of transsexuality may or may not be associated with homosexual behaviour. Transsexuals have been described as women trapped in men's bodies, but some authors have unfortunately and misleadingly linked transsexual behaviour with hermaphrodism and deviancy (see, for example, Gillis, 1986:109). Transsexuals have a strong desire to be *all* female, and may or may not have a bodily type to enhance the female status.

Transsexuals, like transvestites, oversubscribe to the feminine (Hellman et al., 1981), and deal with their sexuality in an almost theatrical way. They are often referred to as 'twilight people', for their ultimate goal is to be heterosexual, yet their sense of comfort and acceptance is within the gay framework. A confusing issue arises as to their sexual behaviour. The negotiation of sex with heterosexual ('straight') men becomes difficult, depending both on their bodily structure and on circumstances. Hence many of them attempt to resolve their sexuality within a clique of transsexuals, and become both homosexual and 'pseudo-lesbian' in their sex activities. It is not uncommon to see transsexuals with gay partners or being intimate with other transsexuals. Of note is the fact that transsexuals may incorporate elements of transvestism into their behaviour profile.

Transvestism

Transvestite behaviour straddles homosexual, bisexual, and heterosexual behaviour. Peter Ackroyd, an authority on 'drag' behaviour and transvestism, defines transvestism as the

> act of cross dressing which is accompanied by fetishistic obsessions ... transvestism is now considered to be primarily a sexual obsession and has, during its long history, often been associated with sacred ritual and with the expression of social or political dissent (1979:10).

Transvestite behaviour within the gay context has been incorporated into the sub-culture as part of its ethos. These transvestites (sometimes referred to as 'drag queens') are homosexual. Cross dressing for them is part of a ritual obsession as well as the creation of an alternative sexual outlet, and on occasion they 'drag' for effect. In Johannesburg, as in Cape Town, one discotheque in particular caters for the sub-culture of 'drag', and beauty competitions are encouraged. Like transsexuals, transvestites are regarded with a certain amount of disdain by many homosexuals. However, because of their overt feminine characteristics, they appeal to a proportion of gay men who succumb to the soft, feminine qualities that they (the transvestites) project. These are often homosexuals who feel trapped in the power struggle of masculine versus feminine issues, and who have not resolved aspects of their identity, so that in responding to the youthful and feminine qualities of transvestites, they are in effect responding to a metaphor of heterosexual behaviour.

Androgyny

The Greek derivation of the term androgyny is from andros (male) and gune (woman). The literal interpretation corresponds to hermaphrodite (half man, half woman). Androgyny, in the contemporary sense, is a reflection of a form of sexual expression and identity. This has occurred chiefly through identification with cults and cult 'worship', and as an ideological response. Androgynous behaviour is often an attempt to defy protocol by creating an image of sexuality that encompasses both male and female nuances (Singer, 1977).

An often-used description of androgyny is that it is a manifestation of *asexual* behaviour. This is a gross error, for asexuality does not exist. Asexual behaviour might attempt to disguise or mask or avoid sexual issues, but the fact that a person says he feels asexual, or dresses in a so-called neutral fashion, paradoxically reflects a sense of sexuality, albeit in a strangely stated way. Androgyny, although diverse in its recent manifestations (through celluloid exposure and sub-clique activities), has direct relevance for homosexual identity and behavioural issues. Firstly, it often manifests in people who are clinically confused, and gives them time and opportunity to experience alternative responses to their identity. Androgyny invites sexual comment, but it is not quite as blatant as full 'drag'. Secondly, it facilitates imaginative dressing and publicly validates a person's need to express male and female components. It is now not uncommon to see men in the streets of Hillbrow, Johannesburg, wearing skirts, leather jackets, and light make-up.

Furthermore, androgyny offers an opportunity to those who 'drag' to tone down their female person, and to reveal a more neutral image of their sexuality. Androgyny, too, has endorsed or even legitimized the concept of 'the third sex', an issue which is closely monitored by leftist and Marxist homosexuals who believe that androgynous behaviour is ideologically correct, for it blurs the traditional stereotype of male dominance and female submission. In essence, for them androgyny is a new uniform of sexuality.

Heterosexual homosexuality (the 'straight gay person')

Writing on male sexuality, but with application to both sexes, Kinsey et al. caution against the danger of attempting too explicit a differentiation between homo- and heterosexuals. They state:

> Males do not represent two discrete populations, homosexual and heterosexual. The world is not to be divided into sheep and goats. Not all things are black nor all things white. It is a fundamental of taxonomy that nature rarely deals with discrete categories, and tries to force facts into separate pigeon holes. The living world is a continuum in each and every one of its aspects. The sooner we learn this concerning sexual behaviour the sooner we shall reach a sound understanding of the realities of sex (1948:19).

Just as the term 'heterosexual panic' has been used to describe distressing heterosexual thoughts and feelings within certain homosexuals (Goldberg, 1984), so also an amount of homosexual expression is found in heterosexuals. Although this expression is often manifest in pseudo-anxiety or anxiety states (Miller, 1978; Ovesey, 1969), in the context of sexual behaviour homosexual heterosexuality is a manifestation of token acceptance by the 'straight' man of homosexual behaviour. Many 'straight' men live out their bi- or homosexual fantasies through direct interaction with gay people. Acceptance of the gay lifestyle, including knowledge of the gay vernacular, is evident. Companionship, bordering on physical intimacy, takes place. Male 'fag hags', or male 'fruit flies' (the counterparts to female heterosexuals who socialize with male gay people) are part of the extended repertoire of the diverse nature of sexuality. Although they ostensibly live their lifestyles according to heterosexual protocols, straight gays, by their symbolic interaction with members of the gay sub-culture, emotionally experience a sense of extended sexuality that would otherwise leave a void in their existence. This form of emotional homosexuality, together with behavioural expressions of acceptance, is an important facet in the understanding of human sexuality that Kinsey et al. have made so clear.

Homosexual identity and self-esteem

A discussion of homosexual identity development would not be complete without the mention of *self-esteem*. Self-esteem has been linked to body satisfaction and body self-image (La Torre and Wendenburg, 1983; Prytula et al., 1979), as well as to aspects which comprise features of an integrated ego (Stricklin, 1974).

Therefore, central to each stage of the writers' model, and in accordance with the hierarchical provisions of homosexual identity development, is the notion that through the forms of social interaction in which they engage, homosexuals seek to establish and maintain a stable identity. This process is linked to the growth of patterns of self-esteem and self-image, which may be defined as the 'selective appraisal of the self which is influenced through interactions with the environment, the most important parts of which are the people with whom the individual comes into contact' (Finch, 1973:20–1).

While studies support the notion that homosexuals are not likely to differ greatly from their heterosexual counterparts in self-esteem variables (Finch, 1973; La Torre and Wendenburg, 1983), Hammersmith and Weinberg conclude that a *commitment* to a 'deviant [sic] identity is positively correlated with significant others' support of that identity, and that those not fully committed to their identity (or who are in flux) have less support and a minimized or distorted image of the self' (1973:77). In other words, positive self-esteem is related to a person's sense of commitment to his sexual identity, and also related to the presence or absence of support from significant others, such as family, friends, and peers within the sub-culture.

As stated in this chapter, virtually all homosexuals are reared in heterosexual families, heterosexual peer groups, and heterosexual educational institutions. Consequently, they grow up with the same stereotypes, moral judgements, and homophobic responses as most others. Therefore stigma, in the context of social relationships, 'produces a distancing between those with the stigma and those without' (Hammersmith, 1987:176). This stigma threatens a homosexual person's self-esteem and his sense of identity by denying him positive social and emotional support.

In this regard, self-esteem, homosexual identity, and crisis are related. Crisis is the result of threat to certain essential attachments which all human beings must make, and the subsequent inability to cope with that threat. As will be revealed in the following chapter, crisis may be precipitated by state or object losses, which, according to Dixon (1979) would include, *inter alia*, the loss or threatened loss of self-esteem, self-concept, or other ego values needed for psychic equilibrium.

In this regard, especially in the context of homosexual identity and self-esteem, the loss *event* is important, as well as the loss *process*. The process concerns the chain of events, set in motion by, and subsequent to the loss (Kreuger, 1983:583). These include a lack of clarity of self-definition and sense of identity, as well as impaired ability in the person to cherish his own self-definition (Mehr, 1983:179).

Therefore, self-esteem is inextricably linked to the development of a homosexual identity and, in terms of the model advanced in this chapter, may be influenced by:

▶ the tripartite experiences of development;
▶ feelings of rage/anger towards the self as a result of dual experiences;
▶ feelings of ambivalence towards the gay community for creating 'double bind' situations;

▶ feelings of internalized stigma and homophobia; and

▶ feelings of uncertainty and/or hostility pertaining to the legal position of homosexuals in South Africa.

In sum, homosexual identity development is an ongoing process which begins in the early formative years, and continues throughout adulthood. This process involves key stages of growth and the person's recognition of his homosexual identity, the timing and achievement of which may vary with each individual. The process is a series of transitions which entail the loss of conditioned identity and the gain of a new self. This *rite de passage* is thus frequently accompanied by states of crisis, which are the focus of the following chapter.

Notes

1. The reader is referred to Erikson's eight stages of identity development, described in Chapter 7 of Erikson, E H *Childhood and Society*, Second Edition, New York: W W Norton, 1963.
2. The writers do not wish to delve into Jungian psychology, but they believe that the concept 'animus-anima' warrants some explanation. The male-female gender according to Jung has an archetypal component (*Eros* and *Logos*), or the masculine and feminine principles which govern psychological functioning. Jung's ideas suggested that masculine and feminine elements are united in human nature, primarily at an unconscious level, whereby a man can live in the feminine part of himself, and a woman in her masculine part. It is Singer (1977), writing in the context of bisexuality and androgyny, who points to the conscious recognition of the masculine and feminine potential in every person.
3. The age of 19 years has been chosen as a specific cut-off point as it corresponds to the legal age for sexual consent between males (in private) in South Africa. However, this note must be read in conjunction with the statutory regulations concerning homosexual behaviour described in Chapter 6 of this book.
4. See Isaacs, G M 'Working with the Male Homosexual Client: Some Major Theoretical Assumptions', Part 1 of an unpublished M.Soc.Sc. dissertation, University of Cape Town, 1979.
5. This term refers to a person who has or is known to have homosexual tendencies but who has not acknowledged them.

6. Details of the notion of 'field dependent' and 'field independent' people are located in the writings of J B Rotter, particularly *Social Learning and Clinical Psychology*, Englewood Cliffs: Prentice-Hall, 1954, and 'Generalized Expectancies for Internal Versus External Control of Reinforcement', *Psychological Monographs: General and Applied*, 80(1), 1966:1–28.

   In the homosexual context, a 'field dependent' person would rely chiefly on the gay sub-culture for sustenance, while a 'field independent' person would have developed the capacity to choose: he would interact with the sub-culture on his own terms and at the same time find fulfilment in the wider culture.

7. See also Bundy, C 'Street Sociology and Pavement Politics: Aspects of Youth and Student Resistance in Cape Town 1985', *Journal of Southern African Studies*, 13(3), April 1987:303–30.

2 Crisis and the growth of homosexual identity

States of crisis are characteristically associated with homosexual identity formation, and indeed crisis is the context within which much homosexual existence occurs. This chapter thus focuses upon the nature of crisis, theories which promote the analysis and understanding of it, and theories which can be used to guide intervention in crisis situations. The thrust of the chapter is to relate crisis and crisis intervention to homosexual identity development, and its emphasis is that a crisis, when appropriately handled, has the inherent potential to become a catalyst for positive personal growth in terms of increased insight and coping capacities.

Antecedents to crisis theory

Sigmund Freud, it is said, demonstrated and applied 'the principle of causality as it applies to psychic determinism' (Bellak and Small, 1965:6), which postulates that acts of human behaviour have their cause or source in the history and experience of the individual. Freud believed the foundation of acceptable human behaviour to be the successful negotiation of the individual's historical legacy — based primarily on the learning experiences developed during infancy and early childhood. The historical journey from infancy to adulthood is highly influenced by *residues* (Vilfredo Parento's term) of past experiences that have developed during the early years, primarily to reduce biological tensions (Ford and Urban, 1963:117). This suggests, following Freud, that causality is operative whether or not the individual is aware of the reason for his behaviour.

Although criticisms of Freud's approach abound, his theories have undoubtedly formed the basis for further reflection and enquiry into human behaviour. The ego theorists, such as Cumming and Cumming (1964), Erikson (1956), and Habermas (1979), although complementing Freudian posits, conclude that Freud neglected the direct study of 'normal' or 'healthy' behaviour. The seminal difference between the ego protagonists and applied Freudian theory is that the former concur that the ego, the accepted basis for dealing with reality, may function and develop from conflictive, as well as from conflict-free situations. The ego has the ability as an 'autonomous' entity to facilitate the individual's adaptation to the environment. In Freud's later, albeit polemical work on religion (*The Future of an Illusion*, 1922), he was able to turn his attention from psychic determinism to the difficulties experienced by individuals in direct relationship to the demands made upon them by civilization (Murphy, 1984:65).

Psychoanalysis, as originally envisaged by Freud, is an approach that attempts to reinstate the individual's psychic properties based on the assumption that the causal connections operate on an unconscious level. Hence, psychoanalytical procedures include interpretation, dream analysis, the assigning of meaning to symbols, hypnosis, and the regression of the person's psychic manifesto to significant periods in early childhood development. When, however, in *The Future of an Illusion* (as cited by Murphy, 1984:65) Freud directs his attention to conflicts that are induced in the individual by demands made on him by 'civilization', he retains the view that such conflicts are psychoanalytically remediable, and appears not to realize that there is an element of contradiction in what he says because 'civilization' can hardly be influenced by the process of psychoanalysis.

It was Caplan, in conjunction with Lindemann in 1946, who established the concept of community mental health, which in turn led to the development of crisis intervention techniques. In opposition to Freudian theory, Caplan believed that all facets that comprise the total emotional *milieu* of the person must be assessed. He states: 'the material, physical and social demands of reality, as well as the needs, instincts and impulses of the individual, must all be considered as important behaviour determinants' (cited in Aguilera and Messick, 1978:6).

It is within Caplan's psycho-social framework, as related to mental health, that *crisis* is examined in this chapter. Caplan (1961) formulated the desirable prerequisites of mental stability. He regards the ego state, its stage of maturity, and the quality of its structure as basic to mental health. The assessment and nature of the ego state is, according to Caplan, determined by

> The capacity of the person to withstand stress and anxiety and to maintain ego equilibrium; the degree of reality recognized and faced in solving problems; and the repertoire of effective coping mechanisms employable by the person in maintaining a balance in his bio-psychological field (Caplan, 1961:37).

However, the tracing of crisis *per se* as a recorded socio-emotional phenomenon anteceded the work of Caplan and his associates. Recognition of crisis (as well as of crisis intervention) emerged formally as the result of two factors: firstly, human responses to the disasters of war, and secondly as a result of natural disasters (Parad, 1965). The experience of an external crisis was perceived as emotional chaos, leading to loss of functioning within the crisis situation. Writing with reference to military crises, Caldwell (1967), Caplan (1961), and Golan and Vashitz (1974) draw attention to a cluster of observable symptoms which identify some of the emotional prerequisites for a crisis to occur:

- battle fatigue, leading to loss of function;
- depression, bordering upon melancholia;
- passive-aggressive behaviour;
- loss of bodily control, resulting in hysterical conversion, often due to panic and fear;
- loss of security and familiar geographical surroundings, resulting in fear of the unknown;
- fear of dying, including the fear of loss of comrades;
- fear of loss of family and friends;
- fear of personality change due to the horrors of war; and
- fear of bodily mutilation.

The responses listed above could indicate 'appropriate' behaviour in the face of life-threatening events. But it was observed that the end result of the crisis usually manifested in *regressive* behaviour in the persons concerned. In addressing the urgency of these crises, military psychiatry dealt with them in terms of the immediacy of the situation, attending to the psychic needs of the soldiers in the 'here and now'. It was discovered that with this approach, progression (proactive behaviour) as opposed to regression (reactive behaviour) occurred, and that with appropriate intervention the crisis could be turned into an opportunity for the expression of emotions, leading to growth.

In combination with this, Lindemann's work on bereavement (the outcome of his observations of the grief and mourning reactions of the survivors of a fire disaster in a Boston nightclub in 1943), led him to conclude that grief was manifest on both a physiological and a psychological level, and that if people were allowed to emancipate themselves from the lost object(s) and participate in 'grief work' — whereby they 'debonded' from the deceased person — they would be in a position to renegotiate future relationships without contamination from the loss situation (Lindemann, 1944).

Lindemann's observations, in conjunction with Caplan's, led to the formation of the Harvard School of Public Health. The outcome of Caplan's activities in the field of crisis research resulted in the thesis that a crisis does not belong within the traditional medical model of disease and illness. It is *an acute situational disorder*, which may occur in healthy persons and, according to Golan, 'may be superimposed on longer-term chronic conditions ... intervention in such cases is confined to the alleviation of the acute situation without attempting to change the basic personality or to deal with the underlying pathology' (Golan, 1978:27).

Crisis theory

Crisis theory recognizes that crises manifest as identifiable sets of emotional and physical responses, occurring chiefly as a result of an external hazard or danger to the person's current status. Perhaps no one in the field of crisis and crisis intervention has surveyed the area as practically as Naomi Golan. In her book, *Treatment in Crisis Situations* (1978), she has systematically unfolded the theoretical and historical assumptions underlying the crisis phenomenon. Germain and Gitterman (1980) describe her model as an encompassing ecological approach. History, case study material, the responses of human scientists and personality theorists, and applied crisis interventions have all been described by her. A notable lacuna in her work, however, is that she does not correlate crisis to the area of human sexuality, and in particular, to the area of homosexual development.

This chapter attempts to fill this gap, and to relate, where appropriate, the concept of crisis to homosexual identity development.

It was Selye who, in 1974, described the *cues* to understanding crisis within the framework of stress. His phases of 'homeostasis' and 'equilibrium' within the context of responses to stress, led him to examine the existence of alarm reactions and resistance, culminating in exhaustion. For him, adaptive and maladaptive responses, as subsequently elaborated upon by Lukton (1974) and Aguilera and Messick (1978), form the basis of understanding crisis.

The description of a crisis profile is well-known and includes the following (Caplan, 1961). First, a feeling of initial tension arouses *habitual* adaptive responses, or coping mechanisms. The tension increases with continued external stimulation, and on-going attempts are made to reduce it with familiar coping strategies. If these are unsuccessful in reducing the tension, an acute phase follows. This acute phase may then be interpreted as 'crisis'. Thus, Caplan's definition of crisis is as follows: 'A crisis occurs when a person faces an obstacle to important life goals that is, for a time, insurmountable through the use of his customary methods of problem solving' (Caplan, 1961:18).

For the purposes of this book, the writers offer another definition of crisis which, although influenced by Caplan's, places a greater emphasis on the dynamics of crisis:

> A crisis is a period (usually short or medium term) of psychosocial disequilibrium in which a traumatic, hazardous, or dangerous experience or event confronts an individual, couple, family, group, or community. This trauma or hazard may be either anticipated or unanticipated. The danger or threat reduces (and in some cases paralyses) the individual's capacity to deal with the situation by using his/her repertoire of familiar coping mechanisms. The crisis represents either an *internal* response to an *external* threat, or is a manifestation of a developmental or transformational, and usually also emotional, process *inside* the individual. A response to crisis is dysfunctional if the level of tension reduction is insufficient and the means used to ward off anxiety are not successful.[1]

Although the word 'crisis' is established in contemporary professional and lay vocabulary, and is understood by people as an event or situation that gives rise to fear, panic, and discomfort, its subtle interpretation rests on a delicate differentiation between individual and generic responses (Jacobson et al., 1968). By *individual* is meant the existential framework within which the problem-crisis-anxiety triad is dealt with by a particular person, while by *generic* is meant the common situations experienced by people in general. These would include *role* crises; *transitional* crises; *developmental* crises; *accidental* crises; and *situational* crises.

Crisis, thus, can be experienced both individually and collectively. However, a generic focus is upon the characteristic type and course of the crisis, rather than on the particular psychodynamics of each individual in crisis. Robertson (1986:27) conceptualizes these latter differences by referring to the 'idiosyncratic nature of crisis', and suggests that each person is unique in regard to circumstances that threaten his peace of mind. Hence it is possible that within a group of people involved in the same situation, some will exhibit a crisis reaction, while others will continue to function adequately.

The upsurge of social scientific interest in the generic manifestations of crisis has led to extensive documentation of it. Expressions of crisis including suicide, family disorganization, psychiatric emergencies, migration, divorce, adolescence, ego stage transitions, physical illness, infirmity brought on by old age, death and dying, sexual dysfunction, corporate stress, and so forth have been documented in a specific academic journal devoted to the crisis phenomenon[2] and in many other volumes (Aguilera and Messick, 1978; Brammer, 1985; Calhoun et al., 1976; Cohen et al., 1983; Dixon, 1979; Duggan, 1984; Erikson, 1963; Golan, 1978; Hoff, 1978; O'Hagan, 1986; Parad, 1965; Slaikeu, 1984; Zimbler, 1979; Zimbler et al., 1985).

A review of this literature, and in particular the work of Golan (1978), suggests that the following common threads run through crisis theory.

There are essentially two major categories of crisis. The first concerns crises of a *situational* nature. These crises are identifiable within a situational context, with clearly defined generic precipitant factors, such as death, illness, or disaster. The second category concerns crises of a *maturational* or *developmental* nature, associated with the stages encountered by the individual in the passage of life, and dependent on developmental components. According to Slaikeu (1984), situational crises contain the spiral of sudden onset, unexpectedness, emergency quality, and danger. Developmental crises, on the other hand, include a series of transitions characterized by the acquisition of certain tasks within the life cycle of the person, although it should be noted that the developmental transitions of adulthood are qualitatively different from those of childhood and adolescence. However, whether in childhood, adolescence, or adulthood, crisis arises when the accomplishment of a task associated with a developmental phase of the life cycle (Erikson, 1963) is disrupted, thwarted, or made excessively difficult.

Having made the distinction between the two major types of crisis, it is suggested that the homosexual developmental profile may contain aspects of *both* developmental (transformational) and situational crises. A clear example is that of 'coming out'. In addition to having to relinquish adolescent responses in favour of acquiring adult responsibility and status, the homosexual person experiences a build-up of responses to external stimuli, which pertain to his self-concept, his sexuality and fantasies, and to his interpretation of the external world's perception of him. This occurs mainly within the boundaries of his state of homosexual being.

Every crisis involves loss, and may be defined by the loss that it brings about (Brammer, 1985; Isaacs, 1979b; Schoenberg et al., 1970). Two kinds of losses were identified by Isaacs and Zimbler (1984): object losses and state losses. *Object* losses relate to the loss of material objects or persons. Included in this category would be divorce or death, or loss of job, home, or property. *State* losses, on the other hand, relate to the loss of a state of being or a state of mind, and include loss of self-esteem, loss of control, loss of self-image, and loss of faith or hope. State losses accompany *every* crisis. They are usually, but not always, accompanied by object losses; for example, the decision to commit suicide is not necessarily precipitated by any object loss.

Before it is possible to realize the potential for growth or gain inherent in any crisis, and thereby to negotiate it successfully, a full exploration, expression, and incorporation of the losses, both object and state, need to be addressed. Both Erikson (1968) and Hirschowitz (1972) stress the positive potential of crisis. They believe that while the person in crisis may adapt to the situation, such adaptation may be in the direction of either growth (resolution) or continued impaired functioning. Erikson (1968) believes that the successful mastery of crisis is essential for the eight-stage epigenetic approach to the life cycle that he postulates, while Hirschowitz (1972) describes five specific responses that reflect poor coping, and which are unlikely to lead to crisis resolution. These responses can be summarized and related to homosexuality:

▶ Excessive denial, withdrawal, retreat, and avoidance can occur. Fantasy may overlay or merge with reality. Because crisis is often interpreted as an internal response to an external stressor, the defence profile — ranging from mature responses to narcissistic ones — becomes a habitual pattern to avoid the anxiety of harsh reality. Within the homosexual framework as described in Chapter 1, the denial of inner feelings is replaced by accumulated resentment and anger towards the 'punishing object'.

▶ Dependency needs can be dealt with by excessive clinging, or by the counter-dependent avoidance of sources of assistance. Because of the accumulation of guilt, and the inability to maintain psycho-sexual congruence, the homosexual development pattern may be imbued with a sense of mistrust. This sense of mistrust can often be carried through to adulthood, and can lead to generalized homophobia. Malyon confirms this, and from his research findings deduces that 'in particular, the empathic antipathy which distinguishes contemporary social attitudes towards homosexuality tends to bias the socialisation process and, in turn, the intrapsychic development of gay men' (Malyon, 1982a:59).

Homosexuals in crisis, as the writers' empirical data will subsequently show, have feared intervention for two principal reasons. Firstly, they have feared that their crises, in general opinion, might be seen as a psychiatric breakdown rather than as a natural response to anxiety and emergency. Secondly, they have been afraid that to acknowledge the state of crisis might indicate failure to significant others (Salzberger-Wittenberg, 1970). This concern is linked to Hirschowitz's next point.

▶ Emotions can often be denied or over-controlled, with eventual disruptive discharge. This is a key feature in the crisis of homosexual existence. The cumulative ingestion of emotions, part of the 'closet' syndrome, may erupt in a manifested crisis response of disproportionate intensity. This is clearly indicated in the multi-problemed dilemma experienced by a client who was HIV antibody positive, and who was interviewed by one of the writers:

The 28-year-old male, who had been tested for the HIV antibody and found to be HIV sero-positive, described the following cumulative crisis profile. He had made three suicide attempts, the first being at the age of 10 when he recognized his homosexual orientation. Subsequently there was divestment of emotional ties with his family, leading to their rejection of him because of his homosexual involvement. A period of unemployment of two years' duration followed, during which time he resorted to labile relationships in order to satisfy his basic need of shelter while at the same time satisfying his homosexual libido. Because of inherent dissatisfaction with his lifestyle, drinking and drugging were used to anaesthetize his sense of discomfort. Ultimately he lost his latest job because of his sero-positive status.

This vignette demonstrates the life script of an individual with an on-going crisis profile because of unresolved and over-controlled defences.

▶ Because energy is required to avoid overt crisis, the rest-work cycle of the individual can be poorly regulated. Attempts to avoid the crisis are made because of the stigma attached to the loss of control which is symbolically linked to crisis. To compound the issue, constant reminders are offered to the person about the negative aspects of homosexuality. In effect, 'coming out', which is part of the identity formation of the homosexual, leads to the re-emergence of many of the intrapsychic conflicts of early childhood and primary adolescence (Malyon, 1982a:61). Psychic manifestations of anger and despair, avoidance and denial, or overcompensatory behaviour may occur.

▶ The individual cannot invoke help, or cannot (always) use it when it is offered (Hirschowitz, 1972:9–10).

The time span for a crisis is traditionally given as from one to six weeks (Aguilera and Messick, 1978; Isaacs, 1987a). This is the acute period of active disorganization. If access to some form of palliative or remedial intervention is unobtainable or unobtained, then adaptation may not be achieved.

Rapoport has examined the interrelating features of a crisis, namely the hazard, the threat, and the inability to respond with adequate coping mechanisms. She shows that the threat is often linked — if not actually, then symbolically — to earlier threats that have led to vulnerability and conflict (Rapoport, 1965:25). Rosenbaum and Beebe support this vital notion. They examine the catalytic properties of the state of crisis which they believe to have the potential of resurrecting old habits as well as evoking new responses. They also warn about the evocative nature of crisis, and say one should not be seduced by the semantic interpretation of the word. Like Golan (1978) and Rapoport (1965, 1970), they extend the meaning of crisis into the arena of stress and predicament, supporting *The Oxford English Dictionary* description of crisis as a 'vitally important or decisive stage in the progress of anything, a turning point' (Rosenbaum and Beebe, 1975:12).

Crisis theory explicitly discards the medical model, which conceptualizes maladaptations and problems in living according to an illness syndrome (Caplan, 1961, 1964; Kaplan, 1968; Slaikeu, 1984). In the context of examining crisis as a growth-promoting experience, it is interesting to note that the literal Greek translation of 'crisis' is a decision, in contrast to the Chinese ideographic interpretation which indicates that a 'crisis' is both a 'danger' and an 'opportunity'. The use of the word danger is not surprising, since this is a concept with which people identify readily. The notion of crisis as an opportunity is more alien, yet it illuminates what is perhaps the most relevant aspect of crisis theory. Robertson (1986), along with authors such as Aguilera and Messick (1968), Fisher et al. (1984), Golan (1978), O'Hagan (1986), and Zimbler (1979), endorses the fact that crises have the inherent potential to become catalysts for personal growth in terms of increased insight and coping capacity.

Disparities in the literature that examines the concept of crisis as a theoretical construct have been partly caused by confusing *the thematic interpretation of crisis* with the *possible outcomes of crisis intervention* (Lukton, 1974). There is no doubt that the phenomenon of crisis, as expounded by ego theorists including Erikson, has been academically and clinically accepted (Malyon, 1982b). There is also no doubt that crisis theory is no longer in search of scholarly verification (Duggan, 1984; Hepworth and Larsen, 1986; Robertson, 1986; Slaikeu, 1984; Snyman, 1987); rather its focus is an abiding concern with understanding crisis in order to inform helpful intervention in crisis situations.

Crisis intervention and the critical stage

Halpern (1973), after conducting a rigidly controlled empirical experiment, concluded that individuals in crisis would exhibit crisis-type behaviour, and would be less defensive than those who were not in crisis. Halpern's study in effect endorsed the operational concepts devised by Caplan (1964), Lindemann (1944), and Parad (1965), and led Golan to synthesize this material into her definitive model of crisis intervention. Golan's model has been used in South Africa by a number of researchers. Writers such as Flisher (1981),

Flisher and Isaacs (1987), Isaacs (1979b), Joffe (1980), Kahn (1978), Robertson (1986), Weber (1975), and Zimbler and Barling (1975) have applied and tested Golan's principles of crisis recognition and her model of practice, and have found them to be valid.

Isaacs (1979b), in using this model to identify and deal with homosexual crises in relation to anxiety, was able to determine the efficacy of the model in the context of both crisis identification and crisis management.

There is ample evidence of a consensus of opinion in the literature as to the nature of the crisis profile. Thus, before proceeding to discuss Golan's model (1978) in detail, it is of importance to identify some of the factors that contribute to the crisis state, and which consequently have implications for intervention.

Some elements of crisis identified by Lindemann (1944) include somatic distress, sense of unreality, guilt and hostile reactions, and loss of patterns of conduct in combination with grief. Lindemann's observations of the inadequate expression of grief included loss of purpose, hostility, alterations of patterns of conduct with significant others, and agitated depression.

The notion of *the sequence of crisis* is also widely acknowledged. Bartolucci and Drayer (1973) and Darbonne (1967) noted the transition of crisis, while the three periods of crisis first identified by Tyhurst (1970) have been incorporated into the work of Aguilera and Messick (1978), Golan (1978), Hirschowitz (1972), and Rosenbaum and Beebe (1975). These areas are a period of impact, a period of recoil, and a post-traumatic phase (Tyhurst, 1970, cited by Rosenbaum and Beebe, 1975:11). Hirschowitz (1972) elaborates upon this sequence by suggesting an arbitrary time period for each phase, and believes that each phase is characterized by a different perspective, emotions, thoughts, and sense of goal orientation. Kubler-Ross (1969) in her seminal work on the crisis of death and dying, offers the phases of denial and isolation, anger, bargaining, depression, and acceptance.

Crisis is based primarily on the paradigm of *loss*, impending loss, and the anticipation of hope and gain (Isaacs, 1979b; Medora and Chesser, 1980). Isaacs (1979b) has demonstrated that with the onset of the crisis state, a profile of manifest anxiety is evident. Part of this anxiety is related to the spiral effect of loss, which includes loss of self-control, of self-image, of the ability to cope, and of the approval of significant others (Puryear, 1979). The clinical profile to emerge from this state of loss includes the following: a sense of bewilderment; a sense of danger or threat to one's very existence, be it psychological or social; a sense of confusion; a sense of impasse; a sense of desperation; apathy; helplessness; urgency; and a sense of discomfort, both physical and psychological. This anxiety can be described as the

> subjective accompaniment of the awareness of loss. Searching for a solution in respect of the loss object, be it a part of the self, a person in a relationship, or a significant other, will elicit [some of the above-mentioned] anxiety [feelings]. Searching, by its very nature, implies the loss or absence of an object, and is an essential component in anxiety (Isaacs, 1979b:7–8).

Associated with the crisis-anxiety dyad is the concept of *risk*, which is central to the understanding of crisis. Dixon (1979) believes that risk may be experienced as a result of tension and apprehension due to the anticipation of danger, and the need to engage in 'trial and error' (risk) behaviour in order to deal with the anxiety and its dangerous components. He quotes Schachel (1959) in this respect:

> The threat of anxiety as a potentiality can be eliminated only by the actual encounter with the dreaded situation or activity, because until we actually meet the situation we do not know whether and how we will be able to live with it, master it, or perish in it, and thus we cannot transform the unknown and new into something knowable and known. Such encounters mean leaving the embeddedness in the familiar, and going forth to an unknown meeting with the world (Schachel, cited by Dixon, 1979:43).

It is, thus, risk which enables the person to deal with the crisis, thereby creating the potential to convert loss into gain. The tradition of experiencing loss has a cultural value attached to it. Loss, usually associated with a tangible object or situation, is seldom afforded the opportunity of being experienced symbolically. For example, within the homosexual context, many persons are unaware of the *collective loss status* to which they are subjected. The loss of heterosexual achieved and ascribed status, including societal support, is but one example. Within the prevailing 'symbolic manifesto', loss is based upon societal approval or disapproval. Thus gay men may not openly experience or show mourning or grief over their loss, because it suggests 'feminine' or 'weak' behaviour. Loss, too, is associated with object or relationship loss experienced during infancy and childhood (Bowlby,

1970; Klein, 1959; Mahler, 1971, 1974; Scharff, 1982; Winnicott, 1965, 1971).

Because of the actual or symbolic experiences of loss accumulated from the past as well as the present, it is difficult for the person in crisis to perceive that the experiencing of loss makes way for gain. Gain is initially perceived as being inaccessible, because of the immediacy and/or severity of the crisis. However, effective identification and therapeutic negotiation of the *critical stage* open the way for potentiating the loss.

The notion of critical stage, presented by Zimbler and Barling (1975), subsequently refined by Zimbler (1979) and Isaacs (1979b), and implemented by Flisher and Isaacs (1987) in a controlled evaluative study on crisis intervention, offers a further dimension to understanding intervention in crisis. A clear distinction is made between the crisis moment (or experience) and the critical stage. The above authors emphasize that the processing of the state and object losses that accompany crisis require, in addition to the classical intervention styles described by Golan and others, the refocusing of intervention strategies to capture the growth potential of the crisis experience.

The reader will note, during the discourse of this book, that the crisis episodes experienced by homosexuals are exacerbated by the notion of weakness and vulnerability or the idea that homosexuality is associated with an illness syndrome, because their expression of crisis is often disallowed in a hostile parent culture. In this regard, Zimbler stresses that the actual crisis moment may be seen as qualitatively different from the ensuing adjustment and reconstruction stages (reintegration) as proposed by Golan, because the emotional content of the crisis moment may have self-destructive, entropic potential (Zimbler, 1979:144).

This shift in the focus of intervention is based on both Carkhuff's (1969) advocacy of action-oriented and facilitative direction by the intervenor, and Small's (1970) concept of the 'propitious moment', which focuses on the vulnerable peak of crisis. In this regard, Zimbler and Barling write:

> We would suggest that the moment of crisis [propitious moment] requires a warm, empathic, supportive approach from the therapist, while the critical stage which follows allows for the more directive, positive strategies usually associated with crisis intervention [action-orientedness] (Zimbler and Barling, 1975:6).

In the light of the above, intervention at the moment of crisis performs an essentially supportive function. Therefore identifying the critical stage, and applying direct and confrontative intervention strategies, constitute a reintegrative function and cognitively assist the person in placing the crisis in a meaningful growth perspective, while at the same time reinforcing his self-worth and self-esteem. The identification and use of the critical stage are therefore concerned with intervention beyond the provision of support and 'holding' the client in an empathic embrace. It requires the intervenor to deal with a period of emergent directionality (Flisher and Isaacs, 1987:41) which includes prioritizing feelings, challenging responsibility through awareness, engaging growth potential, dealing with manifest anxiety, stimulating the cognitive awareness of the client, and making constructive use of contracting, which ultimately promotes the termination process timeously and constructively (Isaacs, 1979b; Zimbler, 1979).

Before leading into a discussion of homosexual crisis in the broader context of Golan's model of crisis intervention, the construct of 'critical stage' is illustrated by a clinical vignette extracted from one of the writers' case records.

A 32-year-old white male sought help soon after he had been physically and sexually assaulted by two men [*the crisis moment*]. After meeting them at a gay pick-up point, he invited them home, where they attacked him, raped him and robbed him of certain possessions.

The first session facilitated the necessary catharsis, and set the stage for intervention. Rapport was established, facts pertaining to the rape were ascertained, and the state of emotional and physical discomfort was evaluated. Acute features of the rape trauma syndrome surfaced during this session, and within the framework of controlled catharsis, empathy, and ventilation, the client was 'given permission' to talk about the trauma. Upon termination of this session, a tentative contract was entered into, drawing attention to the seriousness of the event, and the possibility of the crisis being interpreted by the client as retribution for past misdemeanours [*crisis holding*].

The second session commenced with a résumé of the client's feelings, particularly with respect to the previous session as well as to his coping strategies during the period between sessions. Features of guilt, embarrassment, mild depression, and sexual fantasy cessation dominated this session. In addition, he feared a recurrence of the traumatic episode, and set about isolating himself from others. He expressed relief at being able to share his trauma with the therapist, but at the same time verbalized uncertainty as to the purpose of intervention. The session closed with a summing-up of the process, with the therapist taking note of the client's telephone number, address, and some identifying details [*crisis containment*].

The third session moved towards action-oriented intervention. With risk in mind, the therapist confronted the assault by linking it to the client's recent past. The symbolism of 'cruising' (picking up men for a sexual relationship) was discussed. The notion of intimacy was thus introduced early on in the therapeutic process, and was used to form the basis for contracting for on-going sessions, once the flooding elements of the crisis had been dealt with. A critical issue emerged from this when the client, in response to direct probing about his feelings of being raped, indicated both a hesitant 'enjoyment' of the sexual assault, and the fear of wanting to return to the scene of his initial contact — in order to test out whether he would be violated again. Guilt emanating from this extraordinary desire had in effect rendered him temporarily impotent [The risk or confronting approach combined with the facilitative components of support, constituted the identification of the *critical stage*].

The symbolic link actually triggered off the client's unfinished business with an ex-lover. He expressed anger and disappointment, and stated that if he and his lover were still together, this might not have happened. The stage was now set for focusing the intervention on intimacy, loss, and anger [*engaging growth potential, and incorporation of loss*]. Emerging from this was a poor set of intimate patterns between them, specifically the ex-lover's inability to respond sexually to the client's needs. During the relationship, the client had inwardly desired his lover to be more assertive, and to penetrate him. The link between rape, penetration, and being regarded as a sexually desirable object only surfaced after the rape incident. Once the critical stage had been identified, the emotions of the client were placed into appropriate perspective. Contracting [*therapeutic commitment*] could shift away from the crisis moment and the immediate relief of symptoms, a key technique of crisis intervention, into the areas of identifying coping potential and testing out new directions — emphasizing the growth-promoting features of crisis.

This vignette asserts the value of the risk or action-oriented approach within the crisis framework, as well as affirming the notion of hope, which is discussed later in this chapter.

In order to explore the theme of risk, the writers will illustrate crisis intervention by means of a metaphorical anecdote, with an analysis in parenthesis.

Scenario 1

1.1 An individual is driving a motor car alone from points A to Z. The journey entails a long period through deserted and desert-like terrain. [*This is the event, or series of events that leads to a successful resolution, in this case the completion of the journey.*]

1.2 A puncture occurs. [*This is the crisis. There is a possible threat to the outcome of the journey.*]

1.3 The person, by consulting the car manual and using the tools at his disposal, changes the tyre, resumes his journey, and reaches his destination. [*Crisis resolution has occurred through regulated and appropriate patterns of coping. The anxiety of being stuck was dealt with.*]

Scenario 2

2.1 The spare tyre is discovered to be flat. The person is now alone in unfamiliar terrain, and is stranded. Feelings of despair, concern, and panic occur. [*A crisis or critical event is apparent. A hazard or danger exists — a threat to personal safety (integrity).*]

2.2 The person cannot change the tyre. The area is deserted and nightfall approaches. [*The recognized defence or coping systems are perceived to be exhausted.*]

2.3 The person has to choose between spending the night alone in the car, or stopping a passing motorist, or making alternative plans for survival. [*A risk factor intrudes into the crisis experience. Part of the risk generates previous experiences of either adaptive or maladaptive coping, as well as the risk of choice. The choice is to give in to the crisis (flat tyre) and to succumb to despair and 'disintegration', or to recognize the danger and seek help during the critical period.*]

2.4 The person stops a passing motorist, at the *risk* of being rejected, or otherwise violated, to seek help. [*The person has to identify the crisis to the motorist. Being vague or embarrassed about the state of the tyre might not incur the appropriate response from the passer-by. This involves a further risk factor, in that the motorist is owning to the state of the crisis. The more embarrass-*

ing the crisis, the more difficult it is to weigh up the consequences or risk as opposed to the actual crisis event.]

2.5 Together the motorist and the passer-by decide on a constructive approach to defuse the crisis situation. [*Because the risk factor is exposed, open negotiation about the dilemma can be expedited.*]

2.6 The journey has an anticipated successful ending once help has been accepted. [*Crisis has been resolved.*]

Scenario 3

3.1 If the person did not stop the passer-by, or otherwise seek help, he could 'perish' in the crisis. This is the literal or metaphorical collapse of a person in crisis (e.g. suicide attempt). [*Negative defences or the absence of coping strategies could lead to pathogenic responses. If the person has had bad experiences in past situations, the memory of which is evoked by the present predicament (for example, being stuck before, or being assaulted by a passer-by) it would be more difficult to evoke the risk response.*]

3.2 By asking the passing motorist for help, the person has acknowledged the existence of his crisis, and is prepared to deal with it. [*The validity of the problem (crisis) becomes apparent.*]

Scenario 4

4.1 When the person travels again, he checks to see whether or not his spare tyre is operational. [*Crisis intervention has appealed to the cognitive process, and a proactive learning experience has occurred.*]

4.2 However, on this journey his car radiator bursts, pre-empting another, non-familiar crisis. [*Crisis intervention presumes that the resolution of one crisis does not necessarily preclude the person from experiencing a similar or different crisis. But there should be less anxiety and an improved ability to cope.*]

The foregoing scenarios both illustrate and deal with the ingredients necessary for understanding the crisis concept, and can be applied to assisting the homosexual person in crisis.

Intervention in homosexual crises

The homosexual person, as described in Chapter 1, has a marginal existence. The sociologist Lee has concerned himself with a critical examination of the relationship between cultural marginality and crisis. In his essay on this topic, he equates the marginal state with constant intergroup readjustment, personal crises, and personal mobility. Marginality is assumed as an identity when 'People ... confront a critical situation in which their customary ways of thinking, feeling, and acting fail to meet their demands as they see them' (Lee, 1966:278).

In elaborating, Lee draws attention to the challenge of crisis, which he sees as an urgent and drastic event that precipitates action and reaction. He identifies 'frustration and aggression' as a 'collective response' to crisis. Frustration, he says, results from the blocking of goal-seeking activity which in turn stimulates (passive) aggression which may be aimed at its source (1966:281). Finally, he leaves the reader with hope, for crises in marginal existence, albeit uncomfortable and reaching critical points which Maslow (1962:69) has termed 'peaking experience', have the potential for adaptation and gain.

Whether he likes it or not, the homosexual as a marginal person will experience a crisis of identity. Although his crises will be manifest at individual or generic (or both) levels, the nature and content of his experiences will be shaped by the following:

▶ the ability to recognize the state of crisis and deal with it;
▶ the degree to which the crisis is allowed expression within the parameters of his social existence;
▶ whether or not the individual interprets the crisis state as pathological or as a normal response to a stressful situation;
▶ the extent to which crises are influenced by a homosexual frame of reference. For example, the homosexual who experiences the demise of a relationship will feel composite hurt, pain, or relief. Some, however, may interpret this ending as a homophobic issue, and will then blame homosexuality as the cause of the demise, and not the process of the relationship between two people which has come to an end. (In this regard, it is of interest that nowhere, save in the context of feminist issues, have the writers ever encountered a reference to 'heterosexuality' being blamed for the crisis of dissolution of a heterosexual relationship.)

Miller (1981) points out that the developmental crisis of 'coming out' and gaining a homosexual identity should be regarded as existentially continuous with any other identity crisis, since existential needs are synonymous with the integration of a stable perception of the self. Miller stresses, however, that homosexual existence is denied such validity, in that it

> is formally discontinuous with normative crises because of the non-legitimated direction of [its] aim. Its intentionality is denied expression ... [yet] crisis is the ultimate plea for a set of meanings which articulate effectively within the emergent self (Miller, 1981:2).

May, Angel, and Ellenberger write that within the existential framework of self-determination and the freedom of choice,

> Authentic existence is the modality in which a man assumes the responsibility for his own existence. In order to pass from inauthentic to authentic existence, a man has to suffer the ordeal of despair and existence with its fullest implication: death, nothingness (May, et al., 1958:118).

The ultimate crisis resolution for the homosexual person is to differentiate between his homosexual pain and personal pain. Thus, the crisis of homosexual existence must be understood from a number of perspectives:

▶ The crisis is extended beyond the boundaries of traditional understanding of generic crisis.

▶ The crisis of homosexual existence is separate from conventionally recognized developmental crises.

▶ The experience of normative crisis, whether accompanied by pathogenic or adaptive responses, is often blurred with homosexual (or homophobic) mythologies.

▶ Crises are often perceived as failures, rather than as responses to critical experiences.

▶ The homosexual ethos, including the sub-culture, is in a perpetual state of flux. Strong elements of homophobia, coupled with AIDS, reflect the continuing state of crisis.

The passage or journey of crisis resolution confronting the average homosexual person is dependent on ego tasks of dealing with *suppressed* material as differentiated from *repressed* material. Hollis and Woods draw attention to the idea that the experience of crisis promotes the opportunity of exposing suppressed material, rather than the hidden depths of repressed behaviour (1981:325). This suppressed behaviour (a logical defence against anxiety) has its roots in the following:

▶ early fantasies and a sense of 'difference';
▶ the symbolic process of dealing with male and female internal and external objects;
▶ experimenting with the foregoing, within a fundamentally hostile environment;
▶ fear of processing fantasies, in terms of reality on a physical or sexual level;
▶ experiencing a duality of existence — an external heterosexual framework, and an internal frame of reference of homosexual imagery;
▶ the existential conflict around the notion of choice;
▶ responding to the cues and patterns of stereotypical behaviour associated with the gay sub-culture;
▶ dealing with the emergence of personal integrity, at the expense of loss of heterosexual values and/or support; and
▶ fear of anticipating entry into the gay sub-culture.

These psycho-social influences highlight the crisis context of homosexual existence. The areas may be examined within the stages of crisis identification offered by Golan,[3] which are crucial both to an understanding of homosexual crisis, and to clinical intervention.

An analysis of Golan's five stages in combination with the homosexual matrix is given below, with commentary at both a literal and metaphysical level. In this regard it is important to differentiate between a hazardous *event* and a hazardous *situation*, a distinction which is lacking in Golan's model. An 'event' suggests a clear-cut, identifiable incident, whereas a 'situation' implies a set of circumstances, or even a state of being resulting from such circumstances.

The metaphysical interpretation does not confine itself to the event only i.e. to restoring the person to a state of comfortable homeostasis similar to that prior to the onset of crisis, but attempts to link the crisis to the person's past history. In this respect, the metaphysical approach incorporates both the crisis moment and the critical stage into its diagnostic and treatment repertoire. Thus, the metaphysical exploration of the hazard uncovers the *legacy* of homosexual sensations in the person, and extends the strategy of crisis intervention from a form of emotional first-aid into a means of therapeutic synthesis of the unrealized self. In sum, Golan's model is employed both to address the immediate crisis of the here and now (Isaacs, 1979b:28–47), and to explore unresolved developmental issues, thereby maximizing the growth potential of crisis.

The hazardous event

The hazardous event is a specific event or series of stressful events that have as their origin an internal or external frame of reference. The hazardous event confronts an individual (or dyad or group or community) in a state of relative stability within his psycho-bio-social situation. The hazard, by virtue of its imminent danger or threat, evokes a definite reaction, and marks the start of a change in the ecological balance. The reaction is to the hazard, and not necessarily to the event itself (Caplan, 1964). Translated into an equation, the probability of crisis is reflected as follows:

P crisis = f (hazardous event x exposure x vulnerability).

As an external threat, the hazard may be interpreted as the homophobic response of society towards homosexuality. The subjective realization of homo-erotic motivation is preceded by the introjection of a 'miasmic' anti-homosexual bias (Malyon, 1982a:60). The initial threat (or hazardous situation) with which the homosexual person has to contend is that others assume him to be heterosexual. His maleness is, in a sense, threatening to his homosexuality. Thus the individual's early experience of homeostasis has the potential for disruption, once he has to separate from his presumed heterosexuality. Internal childhood responses to the external threat are primitive, and only manifest themselves, albeit without cerebral maturity, as symbolic responses. This phase of the crisis cycle reflects in early experiences a sense of existential difference and separateness, expressed through withdrawal or puzzlement, and the unique fantasy process in operation at the time, such as cross-dressing, for example.

Kivowitz (1988), writing in the context of object relations, takes separation one step further and links it to the delicate crisis of 'missing', which involves becoming reconciled to the absence or loss of a significant person, object, or state. Kivowitz emphasizes the connection between loss (in the crisis sense), regret, and emotional growth, and says that 'missing, consciously experienced, is a *sine qua non* of good enough self and object constancy' (1988:261–2). If this has not occurred, unfinished business in the form of residual longing for a familiar object or state (e.g. mother, significant other, heterosexual status) persists. The intervenor thus has the task of creating a second chance for the client to experience missing, in order to facilitate change towards personal growth.

The vulnerable state

Becoming vulnerable is the person's *subjective* response to the hazard. He may perceive or experience it

- as a threat to his sense of integrity or autonomy or instinctual needs;
- as a loss of a person or ability or state of being i.e. object or state loss;
- as a challenge to growth, survival, or self-expression.

Threat, in association with loss, is often accompanied by anxiety, depression, anger, mourning, and grief; and challenge by anxiety, fear, hope, and excitement. Ewing, reflecting upon Caplan's (1964) descriptions of the typical course of a crisis, draws attention to the premise that in the vulnerable state, the individual may experience crisis in four distinct phases:

- an initial rise in tension, which evokes usual problem-solving techniques; if these do not work, then
- there is a further rise in tension, with feelings of ineffectuality; then
- the continued rise in tension, coupled with anxiety, elicits emergency problem-solving mechanisms — either the crisis reaches a point of resolution; or
- complicated tension ensues, with functional disorganization occurring (Ewing, 1978:13–14).

As the homosexual person progresses towards cognitive awareness and relative emotional maturity, vulnerability develops as a direct response to the subjective meanings or experiences of the hazard. These responses might be manifest in a series of behaviours and/or defences which are used to ward off the anxiety, such as withdrawal or oversubscribing to heterosexual/familial expectations. The sense of threat, danger, loss, or challenge is sharpened during this time, and results in a state of confusion that is extended beyond the formative years of identity attainment. This period includes the realization that both fantasies and attraction towards same-sex objects are the focus of his internal frame of reference. An exacerbating feature of this period is that the vulnerable state may be present in the individual for many years. Mann (1973) describes this extended vulnerable condition as the inability of the person to master his anxiety, which ultimately influences both his future course in life as well as his adaptive means. Mann in addition universalizes this according to separation and individuation principles, and believes that vulnerable states are collectively based upon the issues of independence versus dependence; activity versus passivity; adequate self-esteem versus diminished or lost self-esteem; and unresolved or delayed grief (Mann, 1973:25).

Therefore successful mastery of vulnerable feelings is ultimately dependent upon the capacity to tolerate and manage effectively both state and object losses (Atkins et al., 1976).

The precipitating factor

The precipitating factor is a *particular* event in the crisis cycle that propels the person into disequilibrium. It is of great significance to the person, and may coincide with the hazardous event. It, too, is deemed crucial in defining the active state of crisis. The precipitant, or trigger, is an event (or series of accumulated events) which reveal(s) the nature of the conflict, for example, a sexual encounter may precipitate the crisis of 'coming out'.

The precipitant is usually linked to the sense of hazard, and the combination of both may give rise to a further state of confusion and/or panic. The *meta response* is reflected in the following illustration. The hazard could be represented by overall discomfort experienced with homosexuality, while the precipitating factor could be the person's experience with a sexual partner for the first time. If the hazard is unresolved, but the experience with sexual intimacy is good, the all-pervading sense of discomfort with homosexuality could overshadow the favourable initial experience. The precipitant thus stimulates the feeling of 'existential difference' which emerges at the critical stage i.e. the difference has now become operationalized — the fantasy exploration, searching behaviour, and physical intimacy with a person of the same sex is *real*. It is also imperative to note that, if the original hazard or hazards have not been identified or placed into perspective, the precipitating factor, as identifying the current crisis state, will be contaminated with unfinished business from the past.

The state of active crisis

Active crisis is an acute state of discomfort, where homeostatic mechanisms have broken down, tension has reached a peak, and disequilibrium has crippled a person's total functioning. Slaikeu believes that active crisis is self-limited and may be followed by a new adaptation which is qualitatively different from the one that preceded the disruption (1984:19–21). This can be achieved partially in the context of the critical stage paradigm, when acceptance and incorporation of loss are realized.

However, if the crisis experiences are perceived to be non-valid within the person's internal frame of reference, they are suppressed, and block qualitative functional changes. The achievement of equilibrium is thereby jeopardized. Successful negotiation of the stage of active crisis is thus dependent upon four factors:

▶ cognitive perception, where the person gains relevant insight into the situation;

▶ management of affect, where the central feelings related to the crisis are accepted and integrated;

▶ the development of thought patterns to seek the truth pertaining to the crisis; these are, in effect, rehearsals for reality; and

▶ capitalization of the critical stage to promote growth.

Within the homosexual context, active crisis usually creates the forum for identifying areas of stress, which are often easily recognized because the person's defences are lowered. These may be reflected in egocentric areas of self-esteem (Dixon, 1979; Gershman, 1983; Teyber, 1988), a fear of acknowledging or revealing the homosexual self, as well as the anticipated course of homosexuality.

The state of active crisis often incorporates strong feelings of internal ambivalence; the lost object (heterosexuality and its symbolic attachments to the parent system) is engulfed by powerful homosexual feelings. This state also evokes the sociocentric features of blame and/or experience, such as those related to the homosexual sub-culture, the non-functionality of homosexual relationships, parental wrath, social stigma, and so forth, as opposed to egocentric features which examine the internal state of homosexual pain. In this regard, Colgan's words are particularly pertinent:

> Clinically, it appears that developmental issues which are left unaddressed ... surface primarily under circumstances of crisis. Until the crisis emerges, men have little motivation for addressing any underlying developmental issues ... Sometimes, a sense of time passing, or the 'mid-life crisis' forms the basis for reconsidering one's intimate connections. Developmental deficits which reappear under crisis form the primary emotional threats to developing an integrated balance of identity and intimacy (Colgan, 1987:114).

Furthermore, a sense of urgency is captured by the state of active crisis, the resolution of which hinges on a choice or decision about an aspect or direction of a person's lifestyle (Hart, 1984:42).

Period of reintegration

The state of active crisis dissipates. Disorganization and tension (including anxiety) tend to subside. Reintegration has an element of future anticipation, which is often linked to hope, resolution, and maintenance (Medora and Chesser, 1980). However, this period might reintroduce anticipatory fears as well, for instance the fear that the crisis will persist, or re-emerge, albeit in a different form. Crisis resolution, as an integral part of the period of reintegration, ought to respond to the ego- and sociocentric features that have been prioritized by the person. For the homosexual, integration normally poses some additional threat, for it carries a spectrum of imminent future hazards. Part of the threat is linked to the pattern of double messages received by the homosexual person, such as the reaffirmation of his self-esteem by significant people (e.g. the crisis intervenor), and the subsequent, if not simultaneous denigration of this process by parents, the media, the law, rumour, and other sources.

A feature of this period may be *idealization* of homosexuality by the client. Hope is increased — hope for perfect homosexual relationships (often modelled on heterosexual partnerships), for political and social reform, for gay rights to become institutionalized, and for a cure for AIDS to be found. More fundamentally, this phase also incorporates elements of general life crises, including fear of abandonment, commitment, intimacy, loneliness, ageing, and reconciliation of past dishonesties. All of these life crises are juxtaposed with an element of homosexual blame, and can reconstitute the cycle of diminished self-esteem, self-oppression, and internalized homophobia (Malyon, 1982a). These are the issues which must be contextually addressed during the critical stage to ensure a favourable therapeutic outcome.

Although Golan (1978) warns that the phases described above are delineated as if they are separate and discrete, they are not mutually exclusive, nor are they necessarily experienced in the linear progression as described. All phases could occur simultaneously, or they could be experienced on an *ad hoc* basis.

Golan's model has particular value as a tool to understanding crisis, and to guiding attempts to intervene therapeutically in crisis situations:

▶ The model, whether used generically to describe the meta-crisis of existence, or to deal with the clinical formulations of crisis for a particular episode in the life space of a person, gives immediate access to understanding the nature of the crisis.

▶ Emphasis on the precipitant aids diagnostic effort, and is the key to understanding the crisis in its totality.

▶ Use of the model, by allowing the person to experience and deal with anxiety and fantasy, specifically from a cumulative point of view, has relevance for defusing the pejorative nature of crisis that many people express, and places crisis and its resolution into a model of health, rather than disease.

In conclusion, crisis understanding and resolution are dependent upon three central features:

▶ Anxiety management, including risk.

▶ Learned responses to previous crisis episodes. In this respect, Baldwin notes that

> the severe stress, unconsciously and symbolically linked with earlier conflicts (or experiences) stirs up fears that are, as a rule, a threat to or actual loss of someone or something essential to instinctual gratification (or survival) (Baldwin, 1979:31).

▶ The ability and *opportunity* to express and experience the emotions related to the crisis within their *contextual situation*. The crisis experience thus has a proactive, rather than a reactive quality.

Notes

1. With special acknowledgement to Dr Allen Zimbler for his help in formulating this definition.

2. *Crisis Intervention. A National Journal for Applied Research and Development of Stress and Crisis Intervention Services*, published quarterly by the Crisis Intervention Institute, New York.

3. For full details regarding Golan's Stages in Crisis Intervention, the reader is referred to the original source: Golan, N *Treatment in Crisis Situations*, New York: The Free Press, 1978. In addition, the dimension of anxiety and critical stage analysis has been added to the model by Isaacs in Part 2 of his unpublished M.Soc.Sc. dissertation, 'Crisis Intervention as a Form of Therapy for Persons with Homosexual Crises', University of Cape Town, 1979.

3 The homosexual sub-culture and homosexual identity

All world civilizations seem to have produced homosexual sub-cultures within themselves. In Western civilization, descriptions of ancient-historical same-sex bonding date back for centuries (Boswell, 1980). Classical Greek writings reflect homosexual emotional and erotic love. Studies conducted in the East and West, including the well-known studies of Ford and Beach (1951) and Kinsey and his associates (1948), have illustrated a diversity of sexual expression within the context of homosexual behaviour.

Western history reflects spasmodic acceptance of homosexual behaviour, dating from the Greek classical period and including catamite rituals in temples, cosmological interpretations of monotheism and polytheism (Hoffman, 1984), and positive recognition of homosexual behaviour within the Napoleonic Code. Nevertheless, onslaughts against the very presence of homosexual behaviour were omnipresent, and are to be found today. Societal disdain has been grounded in three entrenched beliefs. The first of these is that homosexual behaviour always manifests itself in sodomy. The second is that homosexuality defies the laws of procreation. The third represents homosexuals as a subspecies of humankind, characterized by sin. Various meanings have been attributed to homosexual behaviour, and the following pointers illustrate the cultural heritage commonly imposed upon present-day homosexuality:

▶ Homosexuals are seen to be members of a 'third sex', an inversion of the natural process, defying traditional male-female roles.

▶ Homosexuality is thought to involve a man emulating a woman, with the trappings of cross-dressing, effete behaviour, and passive psychological responses.

▶ Homosexual behaviour is thought to be represented by the activities of paedophiles, including the seduction of young boys.

▶ Homosexual behaviour is thought to reflect a culture of narcissism, a desire to perpetuate male beauty in a declining male body through the love of younger, nubile males.

▶ Homosexuality is considered to be predisposed to dandy, effete, and outwardly bizarre behaviour, designed to confuse sexuality with role performance.

Many historical accounts have detailed homosexual traditions (Boswell, 1980; Katz, 1976; Weeks, 1977), but the historical and philosophical dimensions of homosexual tradition are not the subject of this discourse, and for discussion of these the reader is referred to other sources.[1]

Modern gay culture, which is the focus of this chapter, has its origins in a plethora of historical and behavioural complexities, but its newest and deepest root is in the homosexual liberation movement that has swept Europe and North America in the second half of this century. This movement has attempted to define and legitimize homosexual behaviour, and to describe homosexual love in poetry and literature.

Despite this new dimension to modern gay culture, it nevertheless has a central characteristic in common with gay behaviour in past societies: it coexists as *a sub-culture within a parent culture*. Hence, in this chapter definitions and discussions of the meaning of 'culture' are offered, and attention is given to the ingredients that give rise to sub-cultural patterns of behaviour. The salient theme that will emerge from the chapter, linking as it does with preceding chapters on the growth of homosexual identity and the part of crisis in this development, is the power of the homosexual sub-culture over the gay collective.

Culture

What is perhaps the most widely acknowledged definition of culture was offered by Edward B Tylor in 1871. He said that culture is 'that complex whole which includes knowledge, belief, art, morals, law, custom and other capabilities and habits acquired by man as a member of society' (Tylor, cited by Lee, 1966:43). This definition, being broad, poses some difficulty, for it incorporates features which disallow narrow or particular interpretations. For some, culture represents a path or ideology, a sense of the aesthete, a heritage. For others it represents a distinction between ethnicity and religious tradition. In South Africa, for example, culture is reflected *inter alia* by class structure, racial heritage, religious dichotomies, historical relevance, and language.

A definition germane to this book might be formulated as follows:

> Culture is a history of tradition, both accurate and romantic. It is a defined system of folklore, imagery, and experience. Law, custom, and social behaviour impose themselves on private and collective beliefs which ultimately affect or change the attitudes and behaviours of a diffuse community.

South and southern Africa is a melting-pot of cultural heritage, imposed by colonial exploits and acquired from indigenous populations. Anglo-European tradition has mingled with an entrenched and diverse black culture to contribute to a heterogeneous society. A spirit of Nationalist culture, embodied in a minority white population through the powerful influence of the Executive, the Legislature, and the Church, has fragmented the culture to the extent of producing cultural chaos. Africans (comprising different ethnic tribes, religions, and languages), plus Chinese, Indian, Portuguese, Greek, German, Italian, Anglo-Saxon, Jewish, Malay, Afrikaner, and other elements contribute to the cultural existence of our time. Thus, cultural heritage coexists with cultural displacement — a recipe for individual and collective identity confusion.

Culture, society, tradition, and heritage cannot really be separated. To the social scientist, culture is not what the popular notion of art, theatre, music, literature, and other elements of a refined lifestyle suggest. It goes far beyond these. It is a living force, with roots firmly entrenched in meaning, symbol, ritual, image, and message.

Perhaps the popular concept of 'culture' requires elaboration. The eventual spread of literacy, the power of mass communication, and a sense of exploration and travel have created a mass public which is a prerequisite for 'popular culture':

> Mass communications comprise the techniques by which specialised social groups employ technological services (press, radio, films etc.) to disseminate symbolic content to large

heterogeneous and widely dispersed audiences (Bigsby, 1976:19).

The transmission of a society's heritage, its rules and state of thinking, as well as its surveillance of law and custom, becomes accessible to one generation after another. Thus culture is fluid, infectious, and ever-changing. Whatever the manifest expression of diversity, or however complex the cultural heritage and tradition, cultures have the following essential components:

- ▶ they are determined by ritual, historical precedent, and recording;
- ▶ they are human-based;
- ▶ they are dependent on people's relationship with their environment;
- ▶ they are readily transfused from one society to another, or from one sub-group to another;
- ▶ cultures, or parts of cultures, can be assimilated or rejected;
- ▶ cultures perpetuate and/or re-establish the human tradition of existence, search for meaning, and successful mastery of behaviour; and
- ▶ cultures are essential forms of behaviour as well as interpretations of meaning, faith, and ideology.

The homosexual sub-culture

A sub-culture is an arm or part of the wider culture. It cohabits with its parent body and survives within its own style. Centuries of recorded tradition have illustrated the sub-cultural phenomenon. Some examples are royal households, university or college students, prisons, army training units, drug addicts, sex workers, criminal organizations like the Mafia, and so on.

A sub-culture nearly always exists as a marginal or liminal entity, usually separating itself from the mainstream culture because of central tenets of 'different' and sometimes unacceptable behaviour. A sub-culture is thus a shadow of the embracing culture, perpetuating its own norms, behaviour, style, and often its own linguistic structure (Hayes, 1981a; Henley, 1982; Jay and Young, 1978). The following definition of sub-culture in the context of homosexuality is offered by Bronski:

A sub-culture is any group excluded from the dominant culture, either by self-definition or ostracism. The outsider status allows the development of a distinct culture based upon the very characteristics which separate the group from the mainstream. Over time, this culture creates and recreates itself — politically and artistically — along with, as well as in reaction to, the prevailing cultural norms. No counterculture [sic] can define itself independently of the dominant culture. By definition it is distinct, yet there is always the urge, if only for survival's sake, to seek acceptance. Concurrent with this urge, the ruling culture, which perceives non-conformity as threatening, attempts to diffuse the conflict by eradicating the fringe culture, by either extinction or assimilation (Bronski, 1984:7).

The core components of Bronski's definition should be examined, for in effect they offer the ingredients that help us to understand the gay sub-culture:

▶ Coexistence within a larger cultural framework.

▶ Exclusion from the wider culture's tacit policy of acceptance.

▶ A means to 'capture' members by advocating themes or tenets of behavioural expression that are accommodated by the sub-culture.

▶ An expression of desire to challenge the overall culture.

▶ An ability to manifest flexible behaviour in order to gain the wider culture's approval.

Sub-cultures, countercultures, and fringe cultures, although maintaining different strategies and a sense of differing identities, seek to respond from *without* the mainstream culture. Bronski describes this as the 'uneasy symbiotic relationship between mainstream culture and counterculture' (1984:8). Although Bronski, when referring to homosexuality, uses the terms 'counterculture' and 'fringe culture' synonymously with 'sub-culture', the writers believe that attention should be drawn to some distinctions between these concepts which have direct relevance for culture-identity confusion. Although fringe and countercultures have a legitimate function of challenging the *status quo*, these 'cultures' are in effect politicized because they differ from, criticize, or defy majority norms. They are not necessarily gay. Examples of fringe or countercultures include those of punks, freedom fighters, terrorists, religious groups, drag communes, and the 'new wave' description of young people who regard themselves as 'alternative', and who subscribe to some sexual, social, and ideological values which happen to overlap with those of the gay sub-culture. These are expressed in terms of fashion, political comment, and, in some cases, in sexual behaviour.

To complicate the matter even further, the gay sub-culture includes elements of the fringe, alternative, and countercultures. Examples might include the gay left, gay androgyny, and groups of people who disavow the distinction of 'gay' or 'straight', and who think of themselves as 'alternative'. They usually manifest this self-concept through clothing, music, and political feminism. Plummer, in quoting Clarke (1975), refers to this process as 'the diffusion of style', where it loses its symbolic importance and is torn from the group from which it emerged (Plummer, 1981a:208). Plummer hints at this sense of opposition as a crisis for gays, in that the relevance of sub-cultural styles becomes minimized, hence detracting from a consistent identity base.

There are two key factors in the evolution of a gay sub-culture. The first has been the oppression of homosexuals within society, while the second is the concerted effort to produce a form of sexual iconography (Bronski, 1984:10), whereby messages pertaining to homosexual behaviour can be transmitted via signs and codes which allow the like-minded to identify with one another. In this respect, perhaps no greater symbol exists than the pink triangle which identified homosexuals suffering at the hands of the Nazis during Hitler's reign of terror. The pink triangle and the lambda sign (the Greek letter λ) are the universal symbols of gay pride and identity.

The gay sub-culture is the womb in which the gay identity construct is fertilized. Within the stages of identity development, the sub-culture is the means by which the individual can negotiate his identity. Thus the sub-culture is, metaphorically, the procreative factor that gives rise to the gay identity. Although Bronski (1984) suggests that the evolution of a homosexual identity is necessary to the development of a homosexual culture, the writers suggest that the opposite is true as well.

The gay sub-culture is highly institutionalized (Hoffman, 1968; Plummer, 1981a; Read, 1980; Rueda, 1982), and therefore its maintenance is dependent on gay identity acquisition and participation within its boundaries. The aggregate of identity based on homosexual sensibilities, in combination with gay identity constructs, leads to the concept of metaphorical procreation. This life-giving force imbues the individual with:

▶ unconditional acceptance if 'gay behaviour' is manifest;
▶ the provision of sexual expression and outlets;
▶ venues and meeting places;
▶ the provision of gay liberation forums — tempered by the structure of a given society or country;
▶ a means of dealing with gay issues, such as counselling, as well as the promotion of gay identity via cultural campaigns; and
▶ a way of fostering, promoting, and encouraging overt gay behaviour, as well as providing outlets for gay fringe activities such as 'drag', 'high camp', and theatre (Bronski, 1984).

The above reflects the more positive aspects of the sub-culture. However, it should not be forgotten that the institutionalized mores of the sub-culture arose out of oppression. Oppression is often measured in terms of the visibility of the oppressed person. Stigma consists in the external symbols portrayed to others; it is a badge of identity which causes friction within and without the sub-culture. If homosexuality is perceived in terms of the sub-culture, then the sub-culture will become a victim of oppression. Indeed, oppression can become the very essence of the sub-culture, and moreover, it can perpetuate itself from within because of circumstances dealt with in detail later in this chapter.

An associated feature of the gay sub-culture is the *double-bind effect* that it has for its members. In this respect, Littlewood and Lipsedge (1982:41) note that: 'Outsiders are always conscious of a precarious identity. When rejected, they may attempt to reaffirm or rephrase their original identity — hence, "queers" become "gays".' However, the acceptance of the 'gay' label within the framework of the sub-culture does not necessarily denote self-acceptance: many gays feel hesitant, reluctant, and uncomfortable about their association with the 'gay' scene. The sub-culture may therefore be described as 'Urobic' — a system feeding off a menu of oppressive behaviours, including self-oppression (Gonen, 1971; Hetrick and Martin, 1987; Hodges and Hutter, 1974). Cass (1984) likens the gay identity to that of a transparent social identity. She warns that 'individuals can present an image of themselves (i.e. a social identity) that is at odds with the way that they perceive themselves (i.e. their personal identity)'. She adds that

> the concept of the homosexual identity is an unavoidable part of reality, built into the cultural milieu of the present historical period as part of the psychologies of our time. It consists chiefly of 'non-sexual' areas of awareness, such as the consciousness on the part of homosexuals that they constitute a minority and that their social circles are 'gay' (Cass, 1984:21).

The disparity which Cass addresses constitutes the cultural 'double bind' for gay people. It fosters the on-going and perpetual state of crisis in which gays find themselves in relation to the gay subculture.

Another aspect of the homosexual sub-culture is that it fortifies its collective of persons by advocating and maintaining a sense of separateness. It provides role models, styles, outlets for sexual fantasy, and figures of hero worship. In return, the collective endorses this enterprise and the two forces become locked in an effort to maintain equilibrium.

Two schools of thought exist in regard to this vital issue. First, there are those who believe that every attempt should be made to de-emphasize and de-sensationalize the gay mystique and to break down the liminal barrier. This school believes that if gays demystify their life styles and challenge the autonomy of heterosexist philosophy, then gradual assimilation leading to acceptance will occur (Bronski, 1984:13). The opposing school of thought holds that homosexuality will always represent a minority issue to be oppressed by a patriarchal system, and that in order to survive and maintain a collective identity, a separatist and gay identity needs to be maintained (De Cecco, 1981; Harry and DeVall, 1978; Plummer, 1981a).

It is the writers' belief that neither school of thought is 'wrong'; each has a valid position. What is disconcerting is the schism that the two approaches create. This divide creates the essential *metacrisis* of gay existence. While the sub-culture endorses homosexuality, particularly as an alternative form of behaviour, it also perpetuates the sense of difference. However, it should be noted that the achievement of homosexual identity consolidation (Stage 6 of the model proposed in Chapter 1) represents a life style synthesis which implies an ability to transcend this dilemma, because the individual has gained freedom of choice.

The sub-culture itself is in a state of perpetual crisis. This crisis is the result of one fundamental issue. It has promoted, and not without cause, the issue of sexuality as the main reason for its survival — sexuality as interpreted in primitive, instinctual drives that reaffirm the baseline fantasy system. Thus, when sub-culture is extended to gay rights, political issues, and other non-sexual forms of behaviour, the majority of gay people have difficulty in subscribing to those issues which do not overtly maintain or reinforce a sexual framework. Sub-cultural alienation therefore coexists alongside the mainstream of homosexual deliberations. This can be clearly illustrated within the theoretical framework offered earlier for conceptualizing homosexual identity. For instance, some of the issues relevant to this debate include the following:

▶ Homosexuals have a range of self-perceptions that relate to a homosexual identity.

▶ Homosexual identity is a response to an element of control in respect of maintaining a homosexual collective in the form of a sub-culture.

▶ The individual has the ultimate right to choose whether he wishes to acknowledge a gay identity, a homosexual identity, a generic human identity, or all three.

The homosexual sub-culture as it exists today is as diverse and complicated as its heterosexual counterpart. The diversity of the sub-culture has led to the concept of 'culture psychosis', or culture panic. The outcome manifests in homosexual antipathy towards its own network, the 'gay homophobic syndrome'.[2] Homosexual homophobia exists in two contexts. First, there is fear of revealing love for same-sex objects within the parameters of the larger society. This points to the closet syndrome. Fear of being publicly exposed becomes confused with feelings of same-gender attraction. Second, a love-hate relationship with the sub-culture is evidenced. In the empirical study presented later in this book, participants revealed an ambivalence towards the sub-culture. In addition, in their responses to a question designed to elicit their feelings about a local gay organization, respondents displayed a feeling of mistrust, not of the organization itself, but because, as one respondent put it: 'It will be run by a bunch of moffies.'[3] This love-hate phenomenon, indicating deep-seated ambivalence, is the most powerful force contributing to a variety of crises. On the one hand, the person feels a desperate need to belong. On the other, a constant purge of behaviours is evident. The two following vignettes illustrate the love-hate issue:

A 30-year-old professional man sought help from one of the writers in the early hours of the morning. He presented as confused, highly agitated, and bewildered. He had come from a night club in Cape Town where he had spent the evening in awe. This was his first visit to such a club, and on the instigation of his friends he had decided to visit the discotheque. His resultant state of panic revealed his fear of becoming *gay*, an identity which was foreign to his homosexual fantasies. He had witnessed a large number of persons, young, old, female, and male, in different phases of interaction with each another. Men were dancing together, kissing, and rubbing their bodies. They were juxtaposed with individuals sitting alone or looking sad, and those who were inebriated. The scene precipitated his sense of discomfort. The precipitant (trigger) which led to his panic occurred on his departure from the club, when he bumped into a transvestite person of mature age, who was drunk, dishevelled, and 'falling all over the place'. The cumulative effects of the stimuli which had confronted him during the course of the evening led to the ingestion of this person as his own *alter ego*. The person encountered flashed, as it were, on this image, and, in his own words, he feared that 'I will become like that person'.

His crisis and ensuing panic state, coupled with disorientation, reflected a homosexual panic situation with an aversion to his perceptions of the gay scene.

A 28-year-old man was referred to one of the writers by a psychiatrist. On referral, the client was ostensibly suffering from 'identity panic', compounded with episodes of depression and heavy 'binge' drinking. The unfolding of the client's dilemma revealed a man harnessed by his parents' disappointment in him for not being married, coupled with a history of homosexual fantasies dating back to early childhood. He fled from his parents' grasp by leaving the city of his birth in his late teens, hoping to explore his identity in relative anonymity. When this did not work he returned to his city of birth, and to date had not realized his fantasies in any form of intimate contact with others, excepting for masturbation while exposing himself to strangers from the relative safety of his motor car. He believed that his work demanded a 'neutrality of expression', and he felt unable to seek out the most viable outlet for his expression — the gay scene. As a result, he developed contingency plans, and formed friendships with young, heterosexual boys, who responded to him as to a parent. He would capture moments spent with them, such as at the beach, and store these up as a photo-fantasy armamentary to feed his masturbatory cycle. His point of crisis emerged as a consequence of one of his wards getting married, of the massive exposure given to homosexuality by the press and television in respect of AIDS, and of being confronted directly at work by some colleagues who assumed that he was gay.

When asked about his feelings concerning the gay scene, he responded with horror, although his peripheral knowledge of the scene appeared to be sophisticated. He read gay books, saw gay pornographic movies, and was familiar with the jargon. His reservations (perhaps fantasies) about the gay scene reflected a belief that the gay scene would make his homosexuality apparent, and that he would then be labelled as a 'moffie'. He believed that the gay scene would disadvantage him, since his ultimate goal was to find a companion. He felt that the gay scene was too promiscuous. He did not want to become a 'moffie', but desperately wanted to become gay.

This vignette displays vividly the impressions that some people have of the gay scene. Its skewed nature is rooted in a web of unfinished business in respect of the personal identity of the client. The powerful force of 'gay osmosis' through literature and other forms of searching had prevented this client from discovering his own sense of truth. His self-image (a basis for identity), together with his cerebral patterns, had remained distorted. His way of coping was to project all his fears on to the sub-culture, and thus to renegotiate his anxieties in manageable doses. In other words, he tried to make his problem external to himself. He viewed it as being in the sub-culture, and he was unable to accept ultimate responsibility for his problem until the therapeutic contact.

The reciprocal impact of sub-culture and identity can be explored by examining two key components of the sub-culture in depth. These are sexual behaviour, and relationship behaviour.

Sexual behaviour

For gay men, sex, that most powerful implement of attachment and arousal, is also an agent of communion, replacing an often hostile family and even shaping politics. It represents an ecstatic break with years of glances and guises, the furtive past we left behind (Goldstein, cited by Altman, 1986:7).

Homosexual behaviour, social sex roles, sexual orientation, and their links to masculinity and femininity have been defined with differing interpretations (Bem, 1974; Freund and Blanchard, 1984; Freud, 1977; Goode, 1981; Money, 1974, 1977; Money and Erhardt, 1972; Money and Tucker, 1975; Ross, 1983b; Smith, 1983; Taylor, 1983). More recent studies, such as that of Ross (1983c), have applied a stringent methodology to cross-cultural aspects of masculinity and femininity. Findings of these studies show that there is no relationship between femininity and male homosexuality, and that masculinity is inversely related to homosexuality depending on the degree of sex role stereotyping and anti-homosexual attitudes of the society in which the homosexual lives. Taylor (1983:37) explains stereotyping, offering a description of 'pictures in our head that organise our perceptions of the world', but he also reinforces Ross's findings that stereotypical interpretations held by the public towards homosexuals are correlated with conventional sex role attitudes.

Ross (1983b) challenges the inaccurate representation of homosexuals by inviting researchers to concentrate on the *basis* of homosexual attraction rather than on the *gender* of the partner. He isolates a pertinent passage from Gagnon and Simon (1973), who note that

> We have allowed the object choice of the homosexual to dominate our imagery of him ... little is known of the attitudinal and belief systems of homosexuals themselves, or of public expectations of homosexuals and their roles. Cross-culturally, the meaning of the term homosexual is not congruent, nor are the many facets of what is considered masculine or feminine. For that matter, there is some doubt even in the English-language scientific community whether various measures of social sex roles assess similar facets of behaviour or attitudes (Gagnon and Simon, quoted by Ross, 1983b:5).

Homosexual sexuality is an expression of physical love towards another person or people. It culminates in an act of intimate responses leading to bodily (and emotional) exploration. This exploration is governed by fantasy and genital, oral, and physical sensation. The majority of homosexual sex acts (either individually or with others) culminate in heightened sensate arousal leading to complete ejaculation with orgasm (Bell and Weinberg, 1978; Jay and Allen, 1979). As Silverstein and White (1977) say in their book *The Joys of Gay Sex*,[4] gay sexuality must be seen as part of the continuum of human sexuality, *but* it also needs to be placed into perspective regarding the sub-culture.

Gay sexuality, as the literature has made so clear, has usually been mythically described according to dichotomies. These splits range from the 'active' male coupling with the 'passive' male, to the 'butch-femme' stereotype. The image of the homosexual person, well fed by observation of gay tradition including sensational media coverage, includes the promiscuous and transient experiences of gay sex in steam baths and toilets. Two major sets of opinions exist. One is that homosexuals (or gays) respond by emulating male and female sexual roles. The other is that homosexuals (or gays) are riddled with satyriasis. Neither of these opinions is scientifically sound. Bell and Weinberg (1978), Carrier (1977), Jay and Young (1977, 1979), Masters and Johnson (1979), McWhirter and Mattison (1984), Spada (1979), Tyson (1982), Weinberg (1978), and Whitam (1983) all point to the fact that personal sex role *preference* is an important variable in statistically controlled studies of homosexual behaviour.

Particular cultural circumstances can, however, incline people towards specific expressions of sexual behaviour. Carrier, for example, investigated homosexual behaviour among Turks and Mexicans. Mexican males have rigid inserter-insertee roles, while Turks fear the stigmatization that accompanies passive homosexuality. Carrier draws a distinction between lower economic classes in the United States, whose sexual responses are more classically defined, and middle-class Anglo-American males, whose sex roles are not necessarily dependent on the types of sex act performed (Carrier, 1977:53–65). Carrier's observations are endorsed by the above-mentioned investigators, who have shown that sex partner preference is based on macro-cultural influences, private experiences with early bonding processes and, more importantly, the individual's shared and organized reality within the interactive components of the sub-culture. Bell and Weinberg (1978), Masters and Johnson (1979), and Jay and Young (1979) clearly describe these diverse and non-rigid sex role constructs, breaking the myth that *all* homosexual behaviour is dependent on strict social stereotyping. The following vignette illustrates this notion.

> An interview by one of the writers with an African male who came from a family of academics and professionals, and who was himself a professional, revealed a total absence of role confusion. His intellectual concept of homosexuality was unaccept-

able to him from a traditional point of view. His emotional response was that it simply represented for him a sexual truth. He had no experience of the gay sub-culture. His sexual interactions, which he described as diverse, and which ranged from mutual masturbation and fellatio to mutual anal intercourse, were experiences out of the context of role. He still perceived of himself as male and, irrespective of his sexual activities, the traditional roles of the male in traditional African culture, in this case, transcended the individual man's choice of sexual gratification.

The subject of the above cameo was not a part of the gay sub-culture. For those who are part of it, their sexuality and the sub-culture are inextricably intertwined. Of relevance is the sub-culture's spirit of sexual indoctrination, allowing people to believe that they ought to fit into a mould, rather than determine their own sense of sexual priorities.

Sub-cultural constructs of sexuality may be divided into different strata. These include:

▶ *A focus on genital/oral/anal areas* (hence the concept of 'size queens', or penis size obsession).

▶ *A continual searching for the ideal sexual object*, often reflected in clinical terms as 'nympholepsy' — the ecstatic desire to capture the unattainable.

▶ *A continual exploration of the other's sexual features*, at the expense of individual and mutual personal growth.

▶ *A constant sub-cultural reinforcement of sexual predating* (cruising, camping) as a form of validating a sexual personality.

▶ *Making public, within the parameters of gay meeting places, that sex and sexual encounters are valid.* Hence, within the South African context, public cruising (sex-partner hunting) in areas ranging from nude beaches in Cape Town to the Botanical Gardens in Johannesburg, gives the gay person not only the opportunity to seek out sexual partners, but to manifest a 'public image' as well.

▶ *An exploitation of the gay vernacular to incorporate sexual innuendo at both a subliminal and an overt level.* The following is an example of gay vernacular in the local sense, followed by a translation:

Look at that Clora. What a queen. But wada that lunch. It's a picnic basket. I'd love to pomp her up the Ada, but she looks so Dora'd. I suppose she's rent, or maybe even a Priscilla and will only give me a blow-job. My dear, I suppose I'll have to go home tonight and tilly toss-off alone. Moffies are all alike. (*Translation [soliloquy to a friend]: Look at that 'coloured' boy. What a passive-looking (or effeminate) person. But look at his penis [outlined in his pants]. It is huge. I'd like to penetrate him anally, but he looks too drunk. I suppose he is a male prostitute or even a policeman, and he will only give me oral satisfaction. Oh dear, I suppose I'll have to go home alone and masturbate. Gays are all the same.*)

Attention is drawn to the common feature of feminizing the male pronoun with 'she'. This is a generic pattern of referring to all homosexuals, and not only to those who are effeminate by appearance. It seems to represent a form of gay solidarity, and is used by a wide spectrum of the gay population, who also use 'gay' and 'moffie' interchangeably to indicate a homosexual person.

▶ *Dealing with people (or relationships) as transitional objects.* Therefore, part of the crisis of gay sexuality is linked to the experience of *the transitional object*.[5] This object, whether a person or a relationship, receives a high priority for many gays. Because of the earlier described fixation of levels of sexuality, gay persons respond to *intimacy* primarily through sex. This coincides with the continual searching for an ideal sexual object, discussed above. This purpose may, in 'self psychology' terms, be described as the search for the missing self or narcissistic image. The transitional object (the other person) 'lays the foundation for other kinds of activities ... and permits the sublimated expression of powerful emotions to emerge or be actualized' (Cheschier, 1985:228). The transitional object is the object of hope, linked to the anticipation of legitimacy, since the lost object of heterosexuality needs to be replaced by homosexual imagery.

The transitional object syndrome within the gay collective may embrace some or all of the following:

▷ A narcissistic component whereby the validation of the self is obtained and reinforced via sexual/erotic experiences.
▷ A maintenance of the self (or self integrity) by the conscious, and sometimes also unconscious selection of objects like the self. In

this instance, aspects such as bodily type, age, and erotic attributes often feature. In addition, physical similarities are highlighted, and when the transitional object responds to the person's searching needs, an element of 'incorporating the other' is achieved in order to sustain and gratify the unmet developmental needs of sexual identity.

▷ The transitional object is discarded before emotional attachment occurs. This is a defence against disapproval, and an avoidance of abandonment or rejection. Put differently, the transitional object becomes an immature defence mechanism against the anxiety related to intimacy. In clinical terms, the object is rejected before it can reject, and a new object is sought. In this regard intimacy is maintained at the level of erotic (physical) sexuality, reinforcing the notion of the transitory experience being a sexual one only.

While the concept of the transitional object in essence describes the person's searching for aspects of the self that were usually denied during formative years, the clinician should note that there are additional patterns of transience that do not take on the meaning or the intensity of the transitional object. These include sexual 'stroking'; sub-cultural pressures to recognize the transitional object as a valid part of the gay ethos (thus linking the sanctioning of sexual behaviours to acceptance within the gay network); sexual and physical prowess and beauty, based upon competition; sexual addiction; cruising (camping) for recreation or because of peer group pressure; erotophilia; and sanctioned multiple sexual partners within an existing relationship.

Because of the transitional object syndrome, a distorted pattern of intimacy emerges, for it is based on the prerequisites of sexual prowess. If perchance physical sexuality diminishes during the early stages of attachment, it is taken for granted that the relationship is in demise. Psycho-social and emotional aspects of intimacy are minimized. This form of 'over-separation', to use Colgan's term, leads to the avoidance of affectionate behaviour and the resultant denial of emotional needs (Colgan, 1987:102). In order to restore the sense of balance, a *new* sexual object is sought, either within or without the relationship, or once the relationship has been terminated.

According to Colgan, this phenomenon involves forming and maintaining one's identity according to the clinical features of 'over-attachment'. These features share the characteristic of an excessive need for personal and interpersonal emotional regulation. Over-attachment also depends on others for cues which influence behaviour choices (Colgan, 1987:104).

It is, therefore, not uncommon when observing gay men 'camp' one another in bars, clubs, or cruising areas, to note that they might find up to 20 persons desirable, and change their object of desire as frequently as every few minutes or so. Because sexuality has as its priority a sense of reaffirmation or acceptance, sexual conquest is randomly assigned to levels of acceptance by other people of the visible or external self. As a client reported to one of the writers:

> I desperately wanted sexual contact. I spent an evening at the Sea Point wall [a well-known venue for gay cruising]. During the course of the evening, I made contact with seven men. However, each time, after tentatively agreeing to go home with me, they had a change of mind, and abandoned me [in their respective cars] half way home. I repeated this motion until the seventh person in the space of five hours accepted my proposals whereby we negotiated the sex act.

This vignette clearly illustrates the transitional object syndrome. In a period of five hours the client found seven men desirable. The detachment and reattachment from one object to another results in emotional chaos, a sense of resentment, and a reinforcement of brief encounter sexuality as the primary ingredient for self-actualization.

▶ *Dealing with erotica through the medium of pornography, and responding to the sexual cues offered* by discotheques, steam baths (there are two fully operational in Johannesburg, for example), and public toilets in many well-known shopping centres, as well as a multitude of transient, brief and/or clandestine sexual encounters.

▶ *Active searching for that which is gay.* This is reflected in the 'gay bible', *Spartacus — An International Guide for Travellers*, in which South Africa is featured. It contains details of VD clinics, police activity, places offering sexual companionship (specifying 'types' of people), and gives ratings for clubs, discotheques, and restaurants catering for sexual types such as 'leather', 'S and M', 'rough sex', and sex workers. In the local context, specific cruising places such as bars, steam rooms, health spas, hotels, and clubs form the nucleus of the sub-culture. Most gay novices or 'junior tourists' will seek out elements of this phenomenon, with the primary purpose of establishing sexual liaisons.

▶ *Pursuing erotic reading material*, including scholarly overviews of human sexuality generally, or of homosexuality particularly. This includes subscribing to international gay publications such as *Advocate*.

▶ *Participating in gay sexual practices*. Although Masters and Johnson offer a detailed clinical profile of male sexual behaviour, the manual by Silverstein and White, *The Joys of Gay Sex* (1977), in conjunction with *The Gay Health Guide* by Rowan and Gillette (1978), form the most definitive and comprehensive guide to gay sexuality. Isaacs and Miller, writing about gay sexuality in the context of AIDS, delineate a spectrum of homosexual practices. They warn the reader that, in South African law, many of the activities are proscribed by statute. By listing them, attention is drawn to the clinical realities, and in no way is advocacy of these practices implied. The practices listed by Isaacs and Miller (1985:327–8) are:

▷ auto-stimulation, including masturbation, often accompanied by male erotic-sexual fantasies;
▷ rubbing together of the body/penis, with ejaculation and/or orgasm occurring without oral, anal, or manual stimulation;
▷ oral stimulation by one or both partners, one or both consequently ejaculating with orgasm, with or without ingestion of semen;
▷ anal penetration with or without ejaculation into the anal canal;
▷ use of the tongue, not only for kissing, but to insert into and lubricate the anus;
▷ use of artificial devices, including vibrators, usually applied to sensate erotic zones, including the anus;
▷ use of the fingers, including use of the fingers to insert substances into the anus, often as lubrication for anal intercourse or prostate massage;
▷ other sex practices, involving three or more people, sado-masochistic ritual, cross-dressing, etc. (In some cases orgasm with ejaculation can be achieved without partner contact); and
▷ mutual masturbation.

The foregoing discussion of sexual behaviour within the framework of the homosexual sub-culture has a direct bearing on identity growth. When a re-examination is made of the stages of identity formation offered by the writers in Chapter 1, two salient points require clarification and emphasis:

▶ The false notion of promiscuity.
▶ Sexual exploration as an affirmation of identity.

These two issues are interlocked in respect of their ultimate goal, which is to achieve sexual identity and solidarity within a homosexual-gay framework.

'Promiscuity', as defined by the *Concise Oxford Dictionary*, implies sexual behaviour which is unrestricted by marriage, of mixed or disorderly composition, and indiscriminate. Its synonyms of licentious or libertine behaviour suggest free thinking and disregard of rules.

In the context of sexual acting out, homosexual or gay people on the verge of self-discovery need and want a diversity of experience in order to decipher and determine the validity of their homosexual fantasies. Because of the restricted nature of homosexual expression within the confines of their immediate family and community, fantasy and reality are blurred.

Hence, searching behaviour accompanied by sexual testing out is not promiscuity *per se*, but an attempt to negotiate standards of identification and patterns of sexual comfort, and, of most importance, to achieve a means of self-verification. The crisis of exploration situates itself in the overlapping area of sex behaviour and personal validation. Because of the constant emphasis on sexual interaction, the notion of identity synthesis is strongly linked to sexual prosperity. Sexual searching satisfies many psychological components of the gay person. It allows him, with time, to ascertain his own level of comfort with sexual styles, which are based upon need as well as popularity (for example, because of the present AIDS scare, it is popular to masturbate reciprocally, as well as to recognize the condom as a new and real appendage to the penis).

The repetitive return to sex-searching behaviour that confronts most gay people during their phases of identity consolidation is driven by an anticipated fear. Should confirmation of their validity not be sought by constantly 'playing the field', the chances of slipping out of the sub-culture into obscurity are real. This is associated with the sub-culture 'junkie' or 'adrenalin' syndrome, and accounts for the constant need, even if the person is ensconced in a relationship, to seek out possible sexual outlets. This so-called promiscuous pattern is in effect a contingency to ward off isolation, as well as to reaffirm the person's sense of ongoing integrity.

The dilemma here is the exhausting toll on the psyche, as well as the creation within the person of a sense of disharmony, since, paradoxically, the sub-culture frowns on such behaviour. This paradox leads to the classic double bind situation described earlier.

Relationship behaviour

It must be remembered that the visibility of homosexuals is determined by the response to the gay sub-culture, or gay 'scene'. The full spectrum of homosexual diversity is still not clearly evident. Figures estimating the incidence of homosexuality as being from four to 10 per cent of the population should be accepted with discretion (Geddes, 1954; Rueda, 1982), and although Kinsey's estimates are widely accepted and used specifically by gay activists to shape opinion, they too may have need of revision (Rueda, 1982).

In South Africa, the surface of homosexual demography has hardly been scratched. Although a vibrant gay scene exists in the large centres of the country, people interacting with the scene represent only a fraction of the population who could fit into Kinsey's (1948) structure of homosexuality, or who could fall within the definition of homosexuality offered in this book.

Perhaps the true extent of homosexual demography will always elude investigators. One implication of this is that there is incomplete knowledge of the nature of the relationships of homosexuals who do not interact with the gay scene. Furthermore, even where knowledge is available, homosexual relationships are as diverse and longitudinally complicated as those of their heterosexual counterparts.

Despite these complexities, descriptions of homosexual relationships have been offered by many researchers, among them Berger (1982a, 1982b), Berzon (1979), Harry and DeVall (1978), Joubert (1985), McWhirter and Mattison (1984), Ross (1983a), Tripp (1975), Troiden and Goode (1980), Weeks (1985), and Weinberg (1978). Three common factors emerge from these studies:

▶ Homosexual relationships generally emulate relationships of the parent culture.

► Homosexual relationships are a response to a basic human need for intimacy, companionship, and sexual expression.

► Homosexual relationships, in the majority of instances, exist in isolation from mainstream acceptance.

Within the sub-culture, homosexual or gay relationships have a special significance and structure, and are highly institutionalized. Relationships are classified by titles of participants, such as 'lover', 'special friend', 'partner', or 'piece'. Depending on the level of familiarity and intimacy, each relationship is accorded a specific status within the collective. A 'piece' suggests a transitory relationship, while a 'special friend' denotes a person who is special in the context of a romantic and longer-term union. 'Lover', the most common label, has connotations of romance, explicit sexuality, and short- or long-term implications.

Both the collective (the gay sub-culture) and the individuals concerned will respond to a relationship structure depending on the levels of assimilation, participation in, or separateness from, the gay scene. Cotton (1972) refers to this as a form of differentiation in sexual behaviour with particular regard to social activities. Lovers become adjuncts to the person's social set. After a while, if the love relationship becomes no longer legitimate, the person is either taken up into the set or expelled. This leads to the wariness, described by so many gay people, of the social and sexual incest that occurs within the parameters of the sub-culture, and which leads not only to disillusionment, but to the seeking out of new territories.

Hauser (1962) delineated five phases of homosexual relationships, ranging from sex and physical courting to passionate love alongside emotional courting, to nesting and dealing with long-term implications. McWhirter and Mattison (1984) describe identical phases, with the adage that part of a gay relationship deals with a form of releasing.

None of the authors on homosexual relationships mentioned above deal specifically with the aspect of 'contract' in the relationship which, if honestly dealt with, adds a new dimension to it. Open and specific contracting between the participants allows for expression and agreement concerning *sexual preferences*, and for the *sharing of phases of identity development*. No person enters a homosexual relationship knowing exactly what to do. Sexual styles differ from one person to another. Moreover, identity disparity exists in most homosexual relationships, based on previous experiences and responses to personal stigma and oppression. Identity conflict is especially evident in relationships which are negotiated for reasons of 'unfinished business'. Contracting also allows for clarification of roles and expectations within the relationship; for instance, it can stipulate whether the relationship is to be 'open' or 'closed'. This feature deals specifically with the need of both people to explore other forms of sexuality within the relationship, and whether the relationship can withstand intrusion from others (Bell, 1975).

In conclusion, it may be reiterated that the gay sub-culture provides a motivation for and a maintenance of a gay identity. It owes its existence to the fact that it offers some solutions to problems of adjustment shared by a collective of individuals. Conversely, the sub-culture is not the panacea for the identity needs of *all* homosexuals, particularly those who for ideological or psycho-social reasons regard the sub-culture as alien.

This chapter has dealt broadly with 'the homosexual sub-culture', and has focused on the general relationship that exists between the sub-culture and the formation and consolidation of gay identity. However, the nature and shape of a gay sub-culture are not universally constant: sub-cultures differ from place to place in their specific form and structure, and they develop and change in dynamic interaction with their parent culture, political and social circumstances, and local emphases and needs. The following chapter describes the anatomy of a specific homosexual sub-culture, that of the Greater Cape Town area. Its purpose is to trace how a particular sub-culture reflects the generic characteristics of any homosexual sub-culture, and also to show how a particular sub-culture is singularized by interaction with, and response to unique needs and circumstances.

Notes

1. Some major texts which deal with the history of homosexuality include the following: Boswell, J *Christianity, Social Tolerance, and Homosexuality: Gay People in Western Europe from the Beginning of the Christian Era to the Fourteenth Century*, Chicago and London: The University of Chicago Press, 1980; Bullough, V L *Homosexuality: A History from Ancient Greece to Gay Liberation*, New York: The New American Library, 1979; Katz, J *Gay American History: Lesbians and Gay Men in the USA*, New York: Thomas Y Crowell Co., 1976; Lauritsen, J and

Thorstad, D *The Early Homosexual Rights Movement (1864–1935)*, New York: Times Change Press, 1964; Mieli, M *Homosexuality and Liberation: Elements of a Gay Critique* (Translated from the original Italian by D Fernbach), London: Gay Men's Press, 1980; Weeks, J *Coming Out: Homosexual Politics in Britain, from the Nineteenth Century to the Present*, London: Quartet Books, 1977.

2. A comprehensive definition of homophobia is offered by Morin and Garfinkle:

   Any belief system which supports negative myths and stereotypes about homosexual people. More specifically, it can be used to describe (a) belief systems which hold that discrimination on the basis of sexual orientation is justifiable; (b) the use of language or slang, e.g. 'queer' which is offensive to gay people; and/or (c) any belief system which does not value homosexual life-styles equally with heterosexual life-styles (Morin, S F and Garfinkle, E M 'Male Homophobia', 117–29 in J W Chesebro (Ed.) *Gayspeak: Gay Male and Lesbian Communication*, New York: The Pilgrim Press, 1981).

3. 'Moffie' is a common South African term referring to a male homosexual. Its origin is commonly believed to be the Afrikaans word 'mof' (plural 'mowwe') which refers to a cross-breed of indigenous and European cattle. A more romantic origin of the term is advanced by a well-known South African actor who, in an interview with one of the writers, suggested that the word 'moffie' emanated from the vibrant gay culture of Cape Town's District Six, where he grew up as a child. The word, according to him, is a derivation of the French word 'mauve', literally translated into colours of mauves and pinks.

   Although the word 'moffie' was previously almost always used with sarcasm or contempt, it has gradually lost much of its pejorative connotation (Isaacs, 1979b:68), and today it is widely accepted as a non-pejorative indigenous South African term for a gay man.

4. Silverstein's book, *The Joys of Gay Sex: An Intimate Guide for Gay Men to the Pleasures of a Gay Lifestyle*, New York: Crown Publishers, 1977, is banned in South Africa. Copies of the book can be kept in restricted sections of university libraries, and can be consulted there by bona fide scholars and researchers. The

book is, however, freely available in Britain, North America, most European countries, and Australia.

5. The term 'transitional object' is borrowed from Object Relations Theory, and more specifically from Winnicott (1951). In this book, the term is used to describe the relationship between intimacy, sexuality, and identity within homosexual sexuality and the sub-cultural framework.

4 The anatomy of a homosexual sub-culture: the Greater Cape Town area

This chapter deals with a contemporary gay sub-culture in South Africa, that of the Greater Cape Town area. The choice of this region has been determined by the fact that the respondents to the empirical study reported upon later in this book all lived in Cape Town and its surrounds.

The salient aspects to be discussed in the chapter include, among others, the socio-political influences of apartheid on the homosexual infrastructure, the nature of the sub-cultural system as it existed in the Cape Town area at the end of the 1980s (with indications of similarities that may exist throughout the country), and the influence of the sub-culture on identity issues. Finally, an appraisal is made of the sub-culture in respect of some apparent crisis issues.

The Greater Cape Town area

Cape Town is the southernmost metropolis of the Republic of South Africa, and it is also the country's oldest city. It is idealized by many white South Africans as the 'Mother City' of the Republic, so that romanticized descriptions of it abound. The following is an example:

> National sentiment fixes Cape Town as the focal point of South African history and character. White civilization gained its first hazardous foothold on that southern peninsula, and from there it

has spread in successive waves into the great hinterland. The atmosphere and character that time alone can create can be sensed where man has established himself along the sculpturesque lower slopes of Table Mountain (Hanson, cited in Pinnock, 1986:9).

This lyrical quotation contains no reference to the severe geographical and social dislocation experienced by the people of Greater Cape Town since the Second World War, primarily as a result of apartheid social engineering.

Greater Cape Town includes the magisterial districts of Cape Town, Bellville, Goodwood, Kuilsrivier, Simonstown, and Wynberg, and the more outlying areas of Stellenbosch, Strand, Somerset West, Paarl, and Wellington. It includes the central city and city fringe, and embraces residential areas adjacent to the central business district as well as portions of the Atlantic and Indian Ocean sea boards.

Thomas (1986), taking into account different estimates and *de facto* adjustments of census figures, suggests a total population in the region of 2.3 million people in 1985. Of this population, the majority (52.8%) are 'coloured' and Asian persons, 30.4% are whites, and 16.8% are Africans. Africans are the most rapidly growing section of the region's population, since with the relaxation of apartheid influx control, ever-increasing numbers of Africans are migrating to the Greater Cape Town area in search of employment.

Within the population as a whole, as well as within every racial segment of it, the majority are in the 16 to 64 year age group (Thomas, 1986). This age-spread represents the most crucial years for the attainment of a homosexual identity, as discussed earlier in Chapter 1.

Although subjective, a conservative estimate of the proportion of homosexual males within the total male population is in the region of 10% (Kinsey et al., 1948; Bell and Weinberg, 1978; Loraine, 1974). There is no reason to believe that the proportion of homosexual men in the Greater Cape Town area is untypical of this estimate, since cross-cultural research suggests that the proportion is relatively consistent within any society. The work of Whitam (1973), for example, which included studies of homosexual populations in the USA, Guatemala, Brazil, and the Philippines, yielded the following tentative conclusions:

▶ Homosexual people occur in all societies.

- The percentage of homosexual people in all societies seems to be the same, and moreover remains stable over time.
- Homosexual sub-cultures appear in all societies, given sufficient aggregates of communities.

Effects of apartheid

Any contemporary map of Cape Town will depict the extent to which buffering and satellite cluster planning has been carried out, in accord with the uni-racial residential areas enforced in terms of the Group Areas Act and its amendments. Pinnock (1986) draws attention to the degree by which the 'white' mountain suburbs have been separated from the African and 'coloured' settlements of the Cape Flats by a wide stretch of empty land and freeways.

The Group Areas Act was used to declare the inner city 'white', which resulted in the displacement of large numbers of people of colour. District Six, in particular, was one of the areas to disappear, with the population being compulsorily relocated in other areas of the western Cape (The SPP Reports, 1983). By 1970, at least 208 'new' towns for 'coloured' and Asian people had been proclaimed (Pinnock, 1986:17). Furthermore, 150 000 people defined as 'coloured' were displaced and resettled within the Cape Peninsula alone in a period of 20 years.

Pinnock suggests that the remaking of Cape Town slotted in to the urban crises of the 1940s and early 1950s, in order to create worker townships which would provide the minimum needs of labour. He goes on to describe the relocation of people on a massive scale, and says that 'the battle for [white] hegemony was won largely under the banner of "law and order". Cape Town was reshaped and harsh conditions were placed on the rate and conditions of urbanization' (Pinnock, 1986:39).

The physical schism that exists between groups of people who live in symbolic intimacy (e.g. work, labour, and geographical neighbourhoods separated by railway lines, stretches of barren land, or a roadway) has its parallel in a psychological and social framework. The legacy of forced separation has infiltrated the homosexual sub-culture, or parts of it. These parts are affected by an exaggerated sense of separation in various populations. The divisive impact of apartheid on the homosexual collective will be examined in more detail subsequently within the context of the structure of South African society, discussed in Chapter 6.

In recent years, the gradual breaking-down of apartheid has led to an influx of Africans and 'coloureds' to some so-called 'white' areas, such as Woodstock, Observatory, and University Estate. The straddling of people between cultures, or some form of culture fusion, is becoming apparent, but not without consequences. A case example illustrates this particular dilemma:

> A 'coloured' gay man in his twenties, who spent his childhood with his mother, a domestic servant in a 'white' area, trained for a professional vocation at a university. His physical appearance is strikingly 'black', while his psychological and internal frame of reference is strikingly 'European'. His refined accent, style of dress, and achievement in reaching a middle-class standard of living have combined to create a sense of internal and external alienation. He is ridiculed by his 'coloured' peer group and by his family, who believe that he is aspiring to 'white' standards of living, which is seen to be ideologically wrong. The white gay fringe with whom he associates regards his behaviour and attitude set as pretentious and out of place. He is thus culturally, racially, and sexually out of bounds, because of his cross-culture sense of assimilation.

Thus apartheid policy has produced an effect of uncertainty in people's emotional and sexual existence, creating a sense of culture crisis that is not often responsive to remedial intervention. This is therapeutically relevant, in that the above-mentioned client had difficulty in negotiating therapy with one of the writers. The willingness to enter therapy or psycho-social intervention is a culturally controlled or determined phenomenon.

Geographic separation is compounded by the economic and language categories within the population. These divisions are in turn influenced by topography and the history of settlement in Cape Town. For instance, the southern suburbs and the flat-land luxury of areas such as Sea Point and Clifton are inhabited by predominantly English-speaking, upper middle-class whites, who are characterized by a particular political and social flavour, ranging from moderate to liberal expressions of ideology. The northern suburbs, such as Parow, Goodwood, and Bellville, extending to the western Cape Flats, are inhabited largely by Afrikaans-speaking whites with their particular and differing approach to socio-political ideology. Interspersed between these two bastions of white culture are the vast, sprawling 'coloured' areas, including Mitchells Plain, and the African townships, including Guguletu, Nyanga, and Khayelitsha.

Each area has its own sense of class structure, ranging from upper middle-class to squatter camp settlements.

Of paramount importance to understanding the gay sub-culture of Greater Cape Town is that the centre of the city, that is Cape Town itself, is the common meeting place for gay people. Not only does this create difficulties in terms of racial separation, but access to places and facilities (such as the Gay Association of South Africa's 60-10 community centre) is difficult in terms of transport. Gay institutions like bars, clubs, discotheques, restaurants, and late night coffee houses are all located in the Cape Town area. Much of the formal organization of activities is nocturnal, and the implications of access to such activities are far-reaching. Residents of the distant 'coloured' and African areas, unless they have access to motor cars, have to make use of train services, which late at night operate infrequently and which pose a threat to personal safety because of muggings and attacks.

A further paradox is that the 'coloured' population, a set of people with diverse cultural heritage, especially with regard to religion, contributed greatly to the *original* flavour of the gay scene in Cape Town. The local vernacular term 'moffie', now a generic non-pejorative description of a male homosexual, reputedly had its origins within the 'coloured' populace, specifically in the District Six area. Gay dances, drag shows, and Mardi Gras incorporating cross-dressing were part of the Cape flavour in the 1950s and 1960s.[1] With forced removals, the majority of 'coloured' gay people became ghettoed on the Cape Flats. A further reason for the demise of the original 'gay scene' that existed during the 1950s and early 1960s was the architectural alteration of the Foreshore area, including the Victorian central station. Hotels and clubs in the area, specifically 'Skyways' and 'Darryls', were the meeting places for sailors, passengers from ships, local gays, and local sex workers. The decline of the Cape sea route as a result of the re-emergence of the Suez Canal as a major shipping lane, saw the end of gay tourists and a vibrant local gay scene, followed by a gradual relocation of the Cape's 'gay Mecca' (Helm, 1973) to the Transvaal. With the advent of the 'privatization' of clubs on a membership basis (because of alcohol laws and Sunday observance laws, as well as racial separation), clubs, discotheques, and bars became highly selective.

Meeting people across the colour bar was limited to closed parties, clandestine living arrangements, and 'cross-meetings', which usually occurred in insalubrious venues like railway stations and toilets, or else involved travelling long distances to small towns and country areas. In fact one beach on the Atlantic sea board, ironically called Bachelor's Cove, was a favourite day-time cruising area for African and 'coloured' men who wished to liaise with white locals and overseas tourists. Only recently, with the changing of the immorality laws and the opening of beaches, swimming pools, restaurants, cinemas, and other such amenities to all races, have certain clubs and bars, and other gay social resources become available to all clientele. As recently as 1985, police raids on clubs and night spots were undertaken because of racially mixed activities. Now that this has changed, more and more 'coloureds' and Africans are enjoying the facilities offered by gay bars and clubs, but with two sets of consequences.

Firstly, an element of racial integration is apparent, with a steady increase in cross-cultural love and sexual relationships. Secondly, a sense of panic is being expressed by the white gay collective, who feel that people of colour are 'taking over', and who respond by either boycotting certain clubs, or ignoring African and 'coloured' members. This leads to culture chaos, confusion, and anger.

A compounding feature to this sub-cultural predicament is the meta-oppression experienced by *all* homosexuals. The African and 'coloured' gay people who believe that gays are oppressed and should stand together with a collective voice are intimidated by their parent culture or by their heterosexual political counterparts not to join forces with white gays. White gays, who have been suckled on racial prejudice, maintain the *status quo*. They seek out 'coloured' counterparts for sexual interaction, but refuse to extend this into all aspects of egalitarian living. The result of this split has been the emergence of a number of splinter groups which are trying to counterbalance the sense of victimization. Groups such as the Pink Triangle, now defunct, tried to address the needs of the 'coloured' gay community. The Pink Democrats, a national alliance of male and female homosexuals, responded to the needs of racially oppressed people within the context of Marxist philosophy and socialism. Another group, the Organization of Lesbian and Gay Activists (OLGA), liaises with progressive homosexuals from a more political stance. Some religious groups, and groups such as the End Conscription Campaign (ECC), identify in principle with this latter gay group. Actually, these splinter groups are protesting against *the mainstream homosexual network*, which has used the formal organization of the Gay Association of South Africa (GASA) chiefly as a service organization to cater for the specific needs of members. Such service activities include counselling, gay sport organizations, cultural activities, religious groups, and so on. The 'alternative' splinter groups have the primary aim of placing homosexual liberation alongside people liberation. They believe organizations like GASA to be peripheral to political and social reform. This split has had a powerful momentum on the demise of GASA particularly, and is discussed subsequently in more detail in Chapter 6.

Gay social institutions in Cape Town

Read's work, *Other Voices. The Style of a Male Homosexual Tavern*, still furnishes the most adequate account of a gay institution — a gay tavern situated in a downtown section of a city in the Pacific North West of the USA. Read's perceptions of the gay social network are that gays share the same multiple disadvantages that characterize the homosexual sub-culture. These include:

▶ common elements in the structuring of interpersonal relationships;

▶ cognitive and perceptual dimensions of a world view;

▶ lores associated with the 'inside-outside' quality of stigmatized lives (see also Note 4); and

▶ mirror symbolism of the more ritualized behaviours.

Weinberg and Williams (1975) describe gay baths and the social organization of impersonal sex. They examine the organized reality and successful territorial and interactional organization of impersonal sex (a particular link to Read's 'inside-outside' description of the gay social network). They conclude that 'deviance' and its facilitation may be better understood in terms of social organization than disorganization. The availability and popularity of such institutions as gay baths have rapidly diminished in the United States since the advent of AIDS; in Johannesburg, however, some baths and saunas continue to exist.

Gay social institutions in Cape Town may be demarcated into a variety of systems and sub-systems, each being interlinked by need, popularity, accessibility, and sub-cultural values. They may be classified as follows:

▶ gay bars;

▶ gay clubs (including discotheques) and other gay venues;

- gay literature and book shops (or gay sections thereof); and
- gay 'camping' or 'cruising' spots.

Gay bars

At present there are three gay bars operating in Cape Town. On occasions other bars have opened, but have not been successfully patronized. One of the gay bars, located in a hotel in Cape Town, caters mainly for the gay populace. During the day the bar serves drinks and snacks, and from 6 p.m. its activities revolve around the dispensing of alcoholic and soft beverages, as well as providing a large discotheque area. The bar has levels, with a pool room situated at the side. These different areas cater for different sets of particular needs. The upper section provides space for those gays who prefer not to mix with the younger, more vibrant homosexual people whose major aim is to be seen at a homosexual venue. It offers an opportunity for those people who, among other things, may solicit for money, and an area where conversations can be carried out in relative privacy. A definite and more obvious style of camping exists in the one section, whereby cruising rules are less rigid. The lower section, where the music is louder, caters for people who normally socialize in groups of two, three, or larger numbers. It is a place where drinks are had before the patrons venture on to the discotheque, and where people visit after the cinema or a similar outing. There is a distinct difference between the two sections, and the upper section clientele believe the inhabitants of the lower section to be 'prissy'. Patrons on their own prefer the upper section, as it does not carry the obligation to socialize. All sections are connected by interleading doors and/or steps, and there is a constant stream of human traffic. Seeking out potential partners or familiar faces is the pattern of bar-migratory behaviour.

In a study of public homosexual encounters, Blachford quotes Delph who has isolated one major characteristic common to these encounters: the virtual absence of verbal utterances. Delph states that:

> individuals learn to use the special presentations, bodily posturing, gestural cues, the manners and informal (but sanctionable) rules unique to the settings; how the distinctive meanings of space, time and manner (or self) separate the erotic worlds from the conventional ones; how public sexuality produces a metamorphosis of the individuals who partake in it, thus transforming

normal selves into erotic selves (Delph, cited by Blachford, 1981:190–1).

If a person is 'fancied' by another, a ritual occurs of eye contact, bodily gesture, and reciprocal smiles or touching of the genitalia. This is usually verified by trips to the lavatory, whereby negotiations, approval, or verification will be conducted at the urinal. Drinks will then be bought, and a brief 'getting to know you' dialogue takes place. Thereafter plans, either immediate or future, are entered into to pursue the contact. The pool area caters not only for those who primarily enjoy the game of table billiards, but also for lesbians who wish to be seen in a 'masculine' context, as well as for those gays who 'look more butch' than the norm. As the pool room is adjacent to the bar area, it reflects a cross-flow of patrons who survey the atmosphere. The gay bar is patronized on six nights a week, and on Friday or Saturday nights there can be as many as 500 customers.

The other two gay bars, located in the centre of the city, cater for a broad range of gay men and women, and include younger boys and older men who wish to avail themselves of sex worker services, or who like 'a bit of rough trade'.[2] Muggings and/or attacks are known to have occurred at these bars, and as a result they are not always popular with many gay people.

The gay bar as an institution is perhaps the most conventional and popular meeting place for homosexuals. It offers a safe place, as well as providing access to partners, friendships, gay jargon, news of importance, fashion, outlets for AIDS information (including safer sex practices) through pamphlet drops, and the testing out of particular skills and strategies for the 'novice' gay.

Until recently, bars were usually segregated according to gender and race. However, bars in Cape Town are now accessible to, and patronized by, a growing group of lesbians and Africans. There is no doubt that Cape Town bars are more insular than the existing bars in Johannesburg, of which there are seven running simultaneously, always full, and also catering for 'alternative gays'. One bar in Hillbrow draws 'punk'-style gays, androgynous gays, transvestites, and transsexuals. The bar grapevine is notorious within the gay community, and people will learn via friends, acquaintances, and local gay publications about the ethos of each tavern.

Gay bars have a longer life span than gay discotheques. Unless the bar is demolished for development reasons, or the premises taken over by new owners, existing bars continue for a considerable time. As one client mentioned: 'the gay bar in Cape Town has been the most consistent homosexual sensation during my seven-year stay in this country'. The population of the bar is, however, fluid. A 'burn-out' syndrome often occurs when a particular individual feels either that he is too familiar with the general clientele and is therefore no longer a 'new face' warranting attention, or that the atmosphere is too oppressive and verges on the 'meat market' (the need and ability to sexualize people). Such people usually leave the bar scene for months or even years, resurfacing either at the express wish of a friend, or during the summer holiday season, when the bar caters for masses of visitors, local or from abroad.

Gay clubs (including discotheques) and other venues

The gay discotheque, or club, is perhaps the most important venue for gay collectiveness and expression. It provides for two types of psychological processes: the positional model, and the personal model. Day explains the positional model as a process of *imitating* a social role. It is not a personal style. The personal model is developed for the sake of testing out or developing personal attributes (Day, 1981:158). His observations capture the basic yet often subliminal functions of a gay club. It is a human resource that links a person's internal fantasies with the external reality of a variety of role models. For assimilation to occur, the individual embarks on a journey of choice and deliberation, in order to functionalize aspects of identity. Cloning behaviour in respect of dress, style, language, and body (physical) presentation often takes precedence. A portfolio of images is stored by the person, and the club becomes the stage for testing out the sense of ritual. The commonality of the club begins to intrude upon and contribute to the personal sense of style, thus exerting a powerful force on aspects of identity.

Such clubs are usually hidden away in the dark recesses of downtown areas, dismal from the exterior, and without open advertisement of the venue as gay (Schurink, 1986). South African gay clubs have had a history of being situated in dungeon-like rooms, cellars, or on the tops of buildings, thereby symbolically attesting to the fact that they are separate from mainstream entertainment, as well as promoting an anonymous and clandestine way of life. This has recently changed. With the mushrooming of clubs in Johannesburg, Pretoria, Durban, and Cape Town, coupled with the relaxation of police activity (police presence is ostensibly prompted by the illegal sale of alcohol and by drug traffic, rather than by homosexual activities), such venues have become 'visible' and more accessible. One Johannesburg club (now defunct), known as 'After Dark', set the precedent in elevating gay discotheques to a respectable status. It opened in an exclusive shopping centre in the northern suburbs of the city. Gays no longer had to negotiate iron doors and ascend or descend stairs to gain entrance. The club was exposed, as it were, to the general public. It proved to be extremely popular with the Transvaal gay collective.

The gay club is primarily a place or venue where gay people are able to express their sexuality and their varied forms of intimacy without fear or ridicule. Ironically, ridicule in the form of competition, isolation, and fear of abandonment come from within the interactions of the club, and not from without. This form of localized homophobia is a major factor in discotheque psychosis, and will be expanded upon towards the end of this chapter. The majority of gay discotheques in South Africa cater almost exclusively for male clientele. Read describes this phenomenon as follows:

> Contrastively [sic], male homosexuals (particularly whites) are in a kind of activist limbo. As males they inherit a legacy of vested social interest and economic principles, and many of them are as sexist [chauvinistic] as heterosexual men. Unlike women, blacks, Latinos, or native Americans, they have *only* the sexual preference — with its avoidable and unevenly distributed discriminatory consequences — as a possible focus for population-wide unity (Read, 1980:10).

Gagnon and Simon (1967), Mileski and Black (1972), and Tripp (1975) have commented on the fact that the promiscuity and anonymity of male homosexual encounters are either absent or minimal in homosexual relationships between females. The South African situation of male-dominated clubs, however, rests entirely on supply and demand with an element of entrepreneurial expertise attached to it. Clubs that offer a new venue, a sense of differing decor, an escape from the familiar, will attract attention whether or not they are run by lesbians or have a large female patronage. Cape Town currently endorses this fact. The most popular city club is frequented by lesbians, people of colour, and 'alternatives'. So successful was this club that another exclusively male homosexual club, situated in the periphery of the central city, was forced to close as a gay venue and reopen primarily as a heterosexual one.

Choice is another matter. The climate of receptivity, based on cult and a sense of 'ownership' and good experience, determines the popularity or style of a club. Perhaps the most successful club in South Africa, and the longest-running one, situated in a downtown Johannesburg building with a castle-like facade, pays tribute to this fact. The club ruled that lesbians as well as gay men had to be admitted. With time, the club's reputation for non-discriminatory practice became unquestionable. Today it is the most popular club in the Transvaal area, catering for all forms of gender/sexual expression and action. Lesbians, of both the 'butch' and 'femme' types, 'drags', neutral-looking businesspeople in suits, the flamboyant, and Africans and 'coloureds' interact with a fluidity not experienced before in a South African club. A similar feature exists in Cape Town, with one proviso. With only one or, at the most two clubs operating simultaneously, the Cape Town population has less choice. This is an important aspect if one compares Cape Town with Johannesburg. Cape Town gays have to contend with the limited availability of clubs, and this brings about cross-cultural and intra-cultural acceptance. In Johannesburg, where because of the larger captured market some clubs can promote stereotypical fancies or separateness based on ideological, sexual, and cult premises, a degree of separateness can be retained.

The descriptions of Johannesburg and Cape Town are incomplete, in that both exclude reference to areas that are primarily devoted to African homosexual sub-culture. No systematic study has been made of this phenomenon, but Soweto reputedly has 15 small shebeens and music halls exclusively for African gays. In Cape Town, a new group of African gays has recently emerged under the title of the African Gay Association (AGA), and has an active membership of up to 70 persons. A particular club in the centre of Cape Town provides a venue for predominantly 'coloured' gay men and women.

The gay club has the following major characteristics:

▶ It is a meeting place where people of the same gender can enjoy dancing and loud music, and revel in the ethos of the discotheque.

▶ It is a safe and comfortable venue where degrees of intimacy, such as dancing, kissing, touching, and holding hands can be experienced without fear of heterosexual censure.

▶ It is a place where sexuality, at its most powerful in terms of image, body language, and fantasy, can be negotiated.

▶ It is a venue where symbolic catharsis of confession takes place, and where symbolic or ritual ownership of homosexuality is in process. Within the confines of a building, a person vibrantly or publicly becomes gay, even if this is only transient.

▶ The club becomes the 'territory of ownership' for the gay person. Read defines 'territory' as the 'enclosed and physically separated premises and the distinctively patterned behaviours both permitted and expected within them rather than a particular residential location' (Read, 1980:69–70).

▶ The club provides access to sexual partners, and becomes the platform for cruising and camping. HIV and AIDS have added another crisis dimension to the sub-culture's contribution towards identity resolution. Ritual, gay iconography, and camping styles have a direct influence on fulfilling the sexual desires of gay people. In other words, camping and sexual identity forge part of the sexual persona of the homosexual. The club, the bar, as well as the cruising areas are still as active as ever but, because of AIDS, cues pertaining to camping are more subtle and cautious. Gay parties, for example, are invested with gay/AIDS innuendo:

▷ To be seen to be *overtly* camping induces a social risk of ostracism by peers. The person (or camper) becomes a 'risk object'.
▷ Blatant camping at parties or clubs where two people leave together creates an impression of licentious and indiscriminate behaviour which arouses suspicion and mistrust. Whereas previously sexual conquests were regarded with optimum sanction and support, they are, in the context of 'public camping', nowadays treated with suspicion.

Although this latter aspect is frowned upon by segments of the gay populace, it is nevertheless still one of the major functions of a club. Cruising areas are often demarcated by interior design (near the bar, for instance), and lighting. Some clubs, particularly in Johannesburg, have provided alcoves or small rooms where sexual activity (such as mutual masturbation) may take place.[3] Schurink (1986), with reference to observations of a South African gay club, remarks on certain 'social processes' observed. While some of his comments may have been accurate at the time that they were observed (during 1979/80), they may not have the same validity now. However, these social processes need mention. The ones identified by Schurink, with additional comments by the writers, are the following:

▶ *Social* — to convey an element of entertainment and basic enjoyment of dancing and socializing.

▶ *Bopping* — to dance, often to be visible on the dance floor, not only to dance with a partner (usually a familiar friend), but to seduce covertly a person of choice. Many people take to the dance floor alone, not only to suggest an element of independence or to 'rave up' on the dance floor, but to indicate availability for contact to a multitude of others.

▶ *Being camp* — the club provides a platform to be camp or to 'camp it up'. Being camp allows the person to dress up or imitate feminine gestures freely. To 'camp it up' suggests verbal intimacies, or the sudden flash of a wrist, or the use of expletives which deride some unfortunate person.

Schurink (1986:35–6) describes two more 'social processes', which in effect typify the core of clubbing. They are, firstly, to socialize with friends and to pick up a 'piece' for the night; and secondly, to interact with similar 'types' of people according to overt and covert sexual behaviour. This point is important, for people who typecast themselves, say as feminine, usually congregate with similar types. Social clustering occurs, providing an element of safety and familiarity, as well as publicly validating the idiosyncratic sense of identity within the embrace of a gay ethos.

Gay clubs are a response to societal rejection which has caused gays to create a social world of their own. Bronski perceives gay institutions as part of the cult of dispossession, and writes:

> It should come as no surprise that gay men, finding that they are not welcome in this world, invent in their imaginations safer, more congenial places. One of the most common themes in gay writing is the creation of edenic [*sic*] situations, free from the world's hostility ... Imagination is especially threatening to a culture that repressively and rigidly defines gender roles (Bronski, 1984:53).

Quoting Brophy, Bronski argues further that the unlimited imagination is subversive, not only because it is primarily sexual in origin, but because it can provide an alternative vision to the 'real' world. It is this sense of imagination that has led to the establishment of 'other' gay institutions, notably gay cafe society, eating houses, and venues which create a safe and open environment for both gays and gay-sympathetic people.

Cape Town has numerous restaurants owned by gay proprietors. While no eating house exists for gays only, an element of comfort and repartee exists between owner, staff, and clientele at these gay-owned establishments. Specifically, one such late night venue for eating, drinks, and cabaret exists in Sea Point, with customers who are primarily gay or gay-sympathetic. It provides a venue where open affection can be displayed, and a sense of 'gay decorum' exists. Johannesburg, on the other hand, has fused gay with alternative behaviour. Rockey Street in Bellevue, a middle-class suburb in the west of Johannesburg, at one stage developed exclusive, late night eating houses for the benefit of gays, 'alternatives', and 'jet set' avant-garde people. Its popularity was sustained for a number of years until it was eclipsed by regular late night venues. The explanation offered for the demise of this venture in part reflects once again on the *territorial imperative*[4] that gays impose on institutions. The minute that they are 'invaded' by others who have no apparent vested interest in gay sensibilities (in other words, for bonding purposes), gay patronage begins to decline, and newer pastures are sought elsewhere. This point has an important link to the fluctuation of popularity within the discotheque area. Specifically in Cape Town (corresponding with Johannesburg), gay clubs are often short-lived (*Exit*, No. 23, 1987:1), which may be ascribed to a number of factors:

▶ The visible gay collective is of limited size. People feel sexually and socially incestuous when seeing familiar faces all the time. Boredom, as well as fear of intimacy, competition, and isolation set in. People in effect reject or abandon the club before it abandons them.

▶ Political issues, such as the exclusively male character of some institutions, have led to patrons not returning. One Johannesburg bar which refused access to women was quickly boycotted by a number of clients, with the result that a new bar opened, drawing customers away.

▶ Members of alternative fringe cultures, including bisexual people and those who believe that gay clubs offer a tacit acceptance of mood, style, and behaviour, flock to gay venues. 'Outside and inside' polarities, which are seemingly amorphous, have been evoked as a charter of legitimacy for the activist causes of many minority groups, such as 'alternatives', drug users, and others. The gay venues have in many respects opened their doors to others, creating a sense of identity shock for many gay people.

▶ Because of the transient reputation of gays, club owners, who perceive their enterprises as businesses and not as community services, examine the mood of the market, and are quick to change the club into an alternative venue. A club can be exclusively gay one week, and heterosexual the next, as was the case in Cape Town. This analysis is expanded by Bronski's description of sexual and gender arrangements of gay identity. He speaks of gay 'stereotypes' as enabling some homosexuals to create an identity, while at the same time reassuring the heterosexual population, who desperately need gay men and lesbians to be identifiable (Bronski, 1984:79–80). One can take Bronski's observations a step further, in parallel with the concept of fringe culture, marginality, and transience. It is just as important for gay people to be able to identify one another, in order to secure an element of culture procreation as described in the previous chapter. Homosexual visibility within the context of sub-cultural institutions impacts upon the nurturing and potential sustainment of gay identity issues. Any threat from external sources is viewed with hostility.

Gay literature and bookshops

Part of the sub-culture is reflected, particularly in North America, Britain, and parts of Europe, by a formal network of newspapers and periodicals (Bronski, 1984), the most popular and widespread being *Advocate*. South Africa, over the years, has seen attempts to publish material dealing with gay issues. Periodicals dating from as far back as 1975 (*Equus*) to the recent GASA-linked publication *Exit* (formerly known as *Link/Skakel*, and published independently) are retailed at bookshops and gay venues. However, as Bronski (1984:158) cautions: 'There is a dialectic between a self-contained gay culture and the dominant culture, based upon the tension between desire for assimilation and the desire to remain distinct and separate.' In South Africa, this position is aggravated by the fact that publications reach only a minority of gay people. Some have also been banned, lowering their credibility with the authorities. The content of *Exit* is also slated by the gay left wing, as it is seen as harbouring stereotypical images, sexist advertisements and personal column inserts, and *status quo* political attitudes. Thus the consumer market is highly limited. However, certain bookshops in Cape Town and elsewhere now openly display gay publications, both fictional and academic. It is not uncommon, especially over weekends, to see people examining gay books quite openly, and one bookshop in Johannesburg's Hillbrow has become a meeting place for gay people.

Gay camping and cruising spots

There is perhaps no more complex or institutionalized international language or set of symbols than the patterns of homosexual mating behaviour. More commonly known as cruising or camping, this aspect of the sub-culture merits specific attention.

Camping or cruising is underpinned by a formalized, almost ritualistic pattern of interaction. Coupled with the mating or bonding process are a variety of venues and meeting places which provide opportunity for the behaviour.

Camping or cruising may be defined as a conscious or deliberate attempt to negotiate a liaison with another person (or people) for the primary purpose of engaging in some form of sexual activity. It involves self-awareness and the awareness of others, with the object of being noticed, observed, or indulging in generally coquettish behaviour.

Camping behaviour is the constant in the gay sub-culture. At the outset, it must be stressed that the verb *camping* must not be confused with the terms 'camp' or 'camp behaviour'. These have a different meaning. Bronski highlights 'being camp' and 'camp behaviour' as the ability of gay men to re-imagine the world around them by satirizing and diffusing real threats: 'By exaggerating, stylising and remarking what is usually thought to be average or normal, camp creates a world in which the real becomes unreal, the threatening unthreatening' (Bronski, 1984:42).

Because camping or cruising cannot be divorced from the sub-culture in most cases, the camping venue becomes sacred and paramount to gay existence. Besides the conventional meeting places such as bars, clubs, and steam rooms, a variety of other places exists in South Africa, some of them corresponding to overseas counterparts, such as railway stations, shopping centres, youth hostels, and hotels (Stanford, 1981).

Cape Town, specifically, is well-known for its relatively 'open and public' aspects of camping. The beach front stretching from the west end of Sea Point right along the promenade is available 24 hours a day for people who wish to seek out a sexual companion. Sandy Bay, a popular nudist beach of international renown, is accessible daily throughout the year for those people wishing to make homosexual contacts, and areas beyond the vegetation at times abound with homosexual sex activities. Certain railway stations, departmental stores (cloakrooms), and streets in the city centre at night offer 'gay tourists' the chance to make contact.

Perhaps most camping takes place in a particular Sea Point area, the magnet of cruising throughout the year and well-known to travellers from abroad and locally. Known as Graaf's Pool, it is an area concreted off from the beach and adjacent to an enormous rock pool, where men are permitted within the confines of the walled area to sunbathe in the nude. Of symbolic and psychological importance is the geographic vantage point of this venue. There is a concrete pathway, floodlit at night, leading over the beach and rocks to the enclosure. During the day one section is, by unwritten agreement, occupied by heterosexual men, and the other by gay men. Sexual gestures, which may be covert or overt, persist almost all the time, such as surreptitious masturbation. At night, when the area is specifically sought out for camping purposes, there is activity from the late afternoon until the sun rises the next day. Here men walk along the public pathway in bright floodlight. Once within the enclosure, their heads appear over the concrete wall while they indulge in sexual behaviour. It is the spotlight, the overt decision to walk 'over the threshold' in public view, that endows a symbolic status to this particular area. Perhaps the Johannesburg counterpart is the bush and shrubbery portions of part of the Botanical Gardens where gay activity, although relatively hidden, takes place in public.

It is the paradox of camping in public places that promotes a sense of vicarious defiance for homosexuals, and provides them indirectly with some public image. Although harassment does occur, such as spasms of 'gay bashing' and police raids, camping has persisted.

The art of camping is dependent upon levels of experience and assimilation within the sub-culture. Although camping becomes more sophisticated with the increase of self-image and the consolidation of aspects of identity, it represents for the gay person a scenario of style, conquest, sexual release, and testing out behaviour.

Styles of camping include driving around in a motor car, and cruising a particular person or people, with the hope of making contact. Such contact is often dependent on cues, which include nods or gestures from the other party, or 'cat and mouse' chasing in the car to seek out mutual confirmation. Other cues range from following a person once a series of eye contacts has been maintained, to fondling brazenly or exposing the genitals. Some persons will camp up to 20 people in one session, until they receive some form of sexual gratification. It is not the sense of sexual orgasm that becomes the ultimate desire, but rather the fact that a feeling of acknowledgement has occurred. In the language of transactional analysis, the person has been 'stroked'. Camping, furthermore, provides an outlet for clandestine contact. Many of the nocturnal activities assist married men, more mature adults, and bisexual people to indulge in fantasy construction in relative psychological safety.

Plummer similarly regards camping as a legitimate form of sexual expression. He concludes:

> Gay casual sex (pick-ups, cruising and objectification) can be seen as a rejection of this narrow [monogamous] definition of legitimate sex, as it expands its range of possible meanings. It includes seeing sex as a form of recreation, simply a game or hobby or fun. It is divested of all its moral and guilt overtones and is enjoyed as an end in itself (Plummer, 1981a:198).

Tripp has noted that some brief encounters can be marked by an intensity and a closeness unmatched in some long-term relationships, and says: 'Sometimes promiscuity includes surprising elements of affection. Even in fleeting contacts ... affection often develops as a by-product of sexual activity' (Tripp, 1987:146). Commenting on promiscuity, particularly as it has affected the gay population since the advent of AIDS, Altman speaks of the central dilemma facing gay men in the wake of the moralism unleashed upon gays who have sex outside of the traditional relationship. Altman perceives the attack on cruising, camping, and unrestricted sex as misunderstood rhetoric. He states: 'All too often such rhetoric is no more than a way of arguing for conventional moral precepts in the name of health needs' (Altman, 1986:159). Altman's concern, which is shared by the writers and by Pegge (1988a), is not the notion of promiscuity *per se*, but rather that sexual activity should be negotiated within the parameters of safer sex. This issue will be examined in more depth in the next chapter.

The sexual objectification which characterizes camping may, however, also be crisis-inducing, due to the using and discarding of the object once completion of the sexual act is achieved. Plummer, quoting Lee, argues that the gay 'ecosystem' provides the social and physical (geographical) arrangements to facilitate mutual sexual pleasure, without fear of rape, unfulfilled expectations, or high costs (Plummer, 1981a:199). Yet for many gays cruising endorses the sexual fixation of homosexual behaviour at the expense of dealing with its emotional consequences. In contrast to Plummer, Colgan is aware of some of the pitfalls of sexual transience and writes:

> Men who rely on patterns of over-separation will perpetuate their need for distance from affective expressiveness by establishing skills only in sexual performance. Some of these men apparently become fixated with sexual behaviour [sic] as an avenue for satisfying their needs for connection with other males. Others protect themselves from losing by insulating themselves from the possibility of sexual rejection by way of alcohol and other drug abuse (Colgan, 1987:111).

The majority of homosexual clients seen professionally by the writers have expressed, in their case histories, a love-hate approach to camping. This love-hate symbiosis may be interpreted in the following ways:

▶ Camping, which is given impetus by the internal fantasy system described in this book, is possibly the first direct acknowledgement of an emerging gay identity. Manifest anxiety accompanies the thought and action of camping. In many instances camping is undertaken to test out the field, and to check out the response of others. In this way the camper gains reassurance about his needs through the approval of others, without necessarily entering into the realm of physical contact. Camping thus tests out the congruency between fantasy and aspects of reality. The process is linked to 'closet' behaviour and private fantasy experiences, and is often the precursor to gaining access to direct sexual experiences with others.

▶ Camping is the link to the sub-culture. In this respect the messages, iconography, cues, gestures, and overt responses are directly linked to acceptance or rejection. The concern manifest is the apparent *readiness* of the person to engage in direct sexual contact. Often, the desire to camp is overwhelmingly strong in the person, but his internalized homophobia presents a barrier. Thus problems with acceptance and rejection surface, often accompanied by sexual dilemmas (for example impotency) and guilt. Furthermore this stage is also dependent upon skill and technique as well as familiarity with styles of camping, in order for the act to be taken to a meaningful conclusion.

▶ Camping is strongly connected to episodes (short- or long-term) of internalized homophobia. Because of the competitive nature of camping, and the fact that all homosexuals and gays are not at the same stages of identity development, acceptance, and synthesis, the gay scene is given a 'bad' status because it is not fulfilling the immediate needs of the person. This generalization and externalization acts as a strong defence against the fear of intimacy, particularly against the fear of being let down. It must also be noted that camping is essentially linked to companionship. Dealing effectively with loneliness, depression, sadness, and the consequences of relationship dissolution is often negated by the need to camp or to reattach. Many clients seen by the writers have reported that psychological crises in respect of gay intimacy create a heightened sense of libido. In consequence, camping is experienced as a remedy for loss, and sexuality and erotica become the healing components.

▶ Finally, since 1984, with an increasing awareness of HIV/AIDS issues in South Africa, camping and promiscuity have been linked — albeit not always correctly — with HIV infection. Camping has thus taken on a new set of meanings. Whereas pre-AIDS it was an extension of the person's sexual self and a life-giving force, it is now regarded with fear and suspicion. The gay ethos has become associated with a death force, and the camper feels let down by the very activity which he has turned to for survival. The ramifications in this regard for intimacy, mental health, and sexual expression cannot be underestimated.

Camping may be necessary in order to verify aspects of identity which need continuous homosexual endorsement. At the same time, feelings of defilement and incompleteness exist if and when the camping becomes the *only* way of gaining a sense of homosexual identity. Camping often reflects a sense of urgency in this regard. This urgency is evident during the acting out stages of the camping ritual and is often conveyed by non-verbal messages. This urgency often detracts from the sexual intent of camping, and rejection ensues because the campee fears the intensity and emotional needs of the camper.

The gradual encroachment upon the gay sub-culture by fringe or alternative people, who have been seen to penetrate the sacred boundaries of camping, has created another form of crisis. The expressive artefacts and concrete objects used in the sub-culture have acquired the same or similar meanings in the wider culture. No longer are earrings, tight pants, crotch-emphasizing clothes, rings on little fingers, and coloured handkerchiefs in back pockets the prerogative of the gay sub-culture. The process of stylization and iconography, once the hallmark of the gay mystique and identity, has been absorbed into the wider youth culture, thereby making it more difficult for gays to identify with their own kind.

In conclusion, the gay sub-culture in Cape Town, and the gay sub-culture generally, is paradoxical in nature, both sustaining and undermining homosexual identity growth. It is replete with contradictory features: unity and fragmentation; activism and stigma; sexual liberation and self-oppression; intimacy and objectification. Gays thus find themselves in a constant state of multiple dilemma, to which has been added a new crisis — that of HIV infection and AIDS, which are the topics of the following chapter.

Notes

1. Information obtained from a personal interview with a well-known 'coloured' actor and TV artist, who, as a young man, participated in the activities described. However, this flavour has not been altogether lost. A discotheque in Salt River, adjacent to the suburb of Woodstock, on occasions hosts drag competitions for so-called 'coloured' gays. In September 1988, one of the writers attended such a performance. Of note was the number of families of gay participants who offered not only their moral support, but who applied make-up and who helped to costume their sons. A follow-up competition was subsequently held in the Cape Town City Hall. A gala evening was arranged in the form of a 'Miss Gay Universe' pageant. A capacity crowd gathered to watch the non-racial contest.

   The writers believe in the importance of narrating these events, for they contradict some of the more widely held myths that homosexual behaviour is 'forbidden' and 'non-expressed' in the so-called 'coloured' communities.

2. 'Rough trade' refers to male sex workers who assert themselves sexually, and who present an image of being rough, masculine, and unkempt. 'Rough trade' may also be extended into the area of sado-masochistic sex.

3. These activities have, however, decreased significantly since the advent of AIDS in South Africa.

4. A term originally developed by Robert Adrey in his book *The Social Contract*, London: Collins, 1970. He hypothesizes that there are three inborn needs in all higher animals, including humankind. These are identity, stimulation, and security. In order to maintain security, territories (both personal and special), and buffer zones between them, are essential for the maintenance of group integrity. This concept of *territorial space* is pursued further by Edwin Pfuhl in his book *The Deviance Process*, New York: Van Nostrand, 1980. He asserts that territories are multiple i.e. public, home, interactional, and bodily. Within the gay context, the home territory is the place where people have relative freedom of behaviour, a sense of intimacy, and control over the area. In order for this territory to be protected from outsiders, certain ritualized reactions, such as insulation, linguistic collusion, and turf defence are negotiated by the occupants.

5 AIDS: the new homosexual crisis

The nature of AIDS

In the summer of 1981, the Centers for Disease Control in Atlanta alerted the medical world to an unexpected outbreak of Pneumocystis Carinii Pneumonia and Kaposi's Sarcoma in young homosexual men who had no known reason to contract these uncommon diseases (Gong, 1985). The diseases were apparently connected to a novel form of gradual collapse of the immune system. Now known as the Acquired Immune Deficiency Syndrome, or AIDS, the phenomenon has become widely recognized within the past decade, leading to growing international alarm.

AIDS has been described, discussed, and debated in an unparalleled media and scientific blitz (Hastings et al., 1987), and of note is the tendency to regard the illness as largely confined to certain 'at risk' groups (Adler, 1987; Hammett, 1986).

In the developed countries, most of the infected people have been homosexually active men (including bisexuals), intravenous drug abusers, and the recipients of contaminated blood products, particularly people with haemophilia (Gong, 1985). Incidence of the disease has been reported worldwide, and it is manifest pandemically in both males and females in central Africa (Biggar, 1986). The occurrence of AIDS has been increasingly noted in heterosexual partners of all the aforementioned groups, and also in infants born to women in the 'at risk' groupings (Borland et al., 1987).

Originally identified as LAV (Lymphadenopathy Associated Virus) by scientists at the Institut Pasteur in Paris during 1983, the causative virus was independently identified by the National Cancer Institute in the United States as HTLV-III (Human T-lymphotropic virus, type three) (Baum, 1987; Gallo, 1984). A more definitive and internationally accepted acronym for the AIDS virus is 'HIV' (Human Immunodeficiency Virus). It is a rare kind of virus: a retrovirus with a unique method of reproduction and mutation. The etymological basis of the terminology applied to this illness is explained in Chart 5.1.

Chart 5.1 AIDS terminology

AIDS
(Acquired Immune Deficiency Syndrome)

ACQUIRED	SYNDROME
through transactions between people (associated with the environment)	a cluster of symptoms/ signs that characterize a disorder

VIRUS
Viruses are extremely small microorganisms that cannot live outside of living cells. Viruses grow and increase within cells, and then infect further cells. Infection is either halted by the body's immune system, or infection becomes overwhelming and causes death

(Adapted from Bracht, 1978:77)

Some commonly used terms associated with the syndrome induced by the HIV retrovirus are briefly defined and discussed below:

▶ **AIDS** Acquired Immune Deficiency Syndrome is an illness caused by the retrovirus HIV. Its presence may be recognized by one or more opportunistic infections that characterize underlying cellular immuno-deficiency. The name appropriately defines the condition. It is 'acquired' (not inherited or genetic) and is therefore associated with the environment. (An exception might be seen in the infant in utero, who acquires the disease from the infected blood of the mother, but a clear distinction must be made between 'inheritance' via the genotype, and 'inheritance' that the phenotype obtains while in the special environment of the mother's womb.) 'Immune' refers to the body's natural system of defence to combat disease, while 'deficiency' indicates that the system is incomplete or lacking. A 'syndrome' is a group of particular signs and symptoms that occur together, and which characterize a disorder.

▶ **Opportunistic infections** Without normal immune function, the body is vulnerable to assault by many environmental toxins and pathogens. An 'opportunistic' infection is one that does not usually arise in a healthy person with a sound immune system, but which strikes when the immune system is defective.

▶ **ARC (AIDS-Related Complex)** This is a combination of physical symptoms, existing over time, that indicates that a person is infected with HIV. Symptoms include, *inter alia*, fatigue, general malaise, persistent low-grade fevers, weight loss unassociated with diet, dry cough, diarrhoea, night sweats, skin rashes, and swollen or enlarged lymph nodes. The advanced phase of HIV infection has also been termed 'pre-AIDS', or 'lesser-AIDS'.

▶ **HIV antibody sero-positive** Antibodies are proteins in the blood produced by the body in response to the presence of a virus. In the case of HIV, the antibodies appear in the blood from two to 12 weeks (or longer) after infection. People who are antibody sero-positive host the virus in their blood and can transmit the infection to others via the exchange of bodily fluids such as blood and semen.

Although AIDS was first reported in 1981, in the following two years a brief moratorium occurred in the dissemination of information about it. Thereafter, the pandemic proportions of the illness were popularly exposed by *Time* and *Newsweek* magazines. This coincided with the first reported cases of death from AIDS in South Africa. Since then, more research, enquiry, and journalistic reporting has taken place regarding AIDS than any other illness. Watson states that in particular homosexual men 'have been bombarded with more medical information that the average person would receive in a lifetime' (Watson, 1984:559).

Several theories have been devised to assist in the understanding of this disease. The aetiologic models include the immunologic overload theory, the interactive model, and the community impact model (Martin and Vance, 1984). The dominant theme to emerge from these models suggests that the aetiology of AIDS is precipitated by one or more biological pathogens, most likely viral in nature, that produce a state of immunologic incompetence (the primary illness). This in turn leads to the defencelessness of the sufferer against a variety of secondary illnesses (Adler, 1987; Beverley and Sattentau, 1987; Martin and Vance, 1984:1 303). Studies revealing idiosyncratic features of AIDS have described the illness variously as 'an African disease', a 'disease caused by a single viral agent', 'compromised immune systems', and as a 'gay plague' (De Cock, 1984; Johnson and Ho, 1985; Kingsley et al., 1987; Layon et al., 1986; McKusick et al., 1985a; Miller et al., 1986; Pegge, 1988a; Watson, 1984).

Of striking importance is the so-called scientific demarcation of two sets of AIDS patterns: 'Western AIDS', which in effect has infiltrated the active homosexual and bisexual populations; and 'African AIDS', which is recognized as a heterosexual illness, and which confronts a vast number of men and women who interact sexually. This division has wide implications for homophobic attitudes, as well as for a political backlash from African countries which regard the West's attitude as patronizing and tinged with racial anger (Ng'Weno, 1987). Although the two patterns are caused by the same virus, the manifestations of the illness differed initially because the affected population groups were different; but the differences are already disappearing fairly rapidly (Ijsselmuiden et al., 1988). In this regard, the writers view AIDS as a human disease, its nature of transmission being primarily through intimate expressions of sexuality (via bodily fluids, particularly semen), and the virus knowing no boundaries when it comes to gender, sexual identity, race, politics, or geographical location.

This chapter explores AIDS with particular regard to the psychosocial perceptions and experiences of male homosexuals.[1] A perturbing feature to emerge from the literature (Altman, 1986; Shilts, 1987) is the constant attack and counter-attack between homosexuals who believe that AIDS must be seen as a heterosexual disease too, and those who believe that homosexually active people comprise the most significant proportion of the risk population (elsewhere than in central Africa).

Layon and his associates (1986), in a selected review of AIDS in the United States, point to the consistent and stable presentation of AIDS in respect of risk groups, the highest being notably homosexual and bisexual men. Their findings are based on people who had a definitive diagnosis of AIDS, and exclude those people who have not yet developed the illness proper, but who have registered HIV-positive on the AIDS test. Their breakdown of AIDS sufferers according to risk groups is as follows:

Gay/bisexual men	72%
IV drug abusers	17%
Haitians	5%
Haemophiliacs	1%

Their fifth risk group is identified as people with 'miscellaneous' contacts, of which between nine and 15 per cent are heterosexual contacts, including women, who have either been the recipients of blood products or who have had sexual relationships with an infected partner. Most AIDS patients range in age from 20 to 49 years, with a median age of 35 years for gay men (Layon et al., 1986:825).

Complementing the above findings, Kingsley, in association with 11 researchers (1987), reported in *The Lancet* the enormity of risk factors in the so-called unsafe sexual practices of homosexual men. The study examined 2 507 homosexual men who registered sero-negative on the HIV test at enrolment, and who were monitored for 12 months to elucidate risk factors associated with sero-conversion to HIV-positive. About 11% of the sample who, *inter alia,* practised recipient anal intercourse with two or more partners during the period, sero-converted. Receptive anal intercourse accounted for nearly all new HIV infections among the homosexual men covered by the study (Kingsley et al., 1987:345).

Isaacs and Miller (1985:327) have alluded to the perhaps understandable but nonetheless unacceptable comment that AIDS research has enjoyed 'disproportionate effort and publication space'. To date, the bulk of scientific reportage has focused on the epidemiology, aetiology, pathogenesis, definitions, and palliative treatment of AIDS. The crisis proportions of this disease, however, have other aspects that have tended to escape concentrated attention. Only isolated attempts have been made to identify and document the AIDS scare-panic syndrome, the characteristics of risk factors, psychiatric depression and neurological factors, counter-transference, and the emotional responses of the 'worried well' (Holland and Tross, 1987; Isaacs, 1987a). While the psycho-social implications of AIDS, and the advocacy of safer sex practices for people at risk have not been ignored, nevertheless writers on AIDS have in general failed to examine the principles of crisis and crisis theory that are called for in dealing with this phenomenon.

AIDS and the homosexual sub-culture

The significance of AIDS and its associated crises for the homosexual sub-culture are now examined with particular attention to the following:

▶ distinguishing features of AIDS and ARC;

▶ the concept of the 'worried well';

▶ the socio-dynamics of a minority population at risk;

▶ the concepts of sexuality and intimacy, and the dangers of metaphorical comparisons;[2] and

▶ AIDS and the development of homosexual identity.

Homosexual sexual practices are not 'new'. Legislation against sodomy (anal penetration) has existed for centuries. Even where legislation against homosexual practices has been revised, such as the state of Wisconsin in the United States, or the Australian state of New South Wales, criminal sanctions against sodomy remain (Altman, 1986:13). The sudden presentation of homosexual sexual acts as a primary cause of the spread of the disease must be treated with sensitive circumspection. Martin and Vance indicate clearly that

> Sexual behaviour must be conceptualized more fully as consisting of several but distinct correlated dimensions, each of which must be measured separately. The most significant dimensions are the number of partners, number of acts, types of specific sexual acts engaged in, characteristics of partners, and the place or location of the sexual contacts (Martin and Vance, 1984:1 304).

The combination of a new mutant virus of possible African origin with the fact that the majority of AIDS-related patients have experienced a multiplicity of sexual partners, points to risk factors within the homosexual population. Sexual contact is currently regarded as the primary mode of AIDS transmission among homosexual men. AIDS itself, and the AIDS-related complex, present a unique constellation of disease characteristics. These are summarized below:

▶ The aetiology of the illness is speculative.

▶ To date, treatment programmes for the secondary illnesses (Pneumocystis Carinii Pneumonia, and Kaposi's Sarcoma) are essentially palliative and ineffectual.

- The incubation period, levels of infection, and mode of transmission of the disease by so-called carriers, are relatively unknown; they are subject to much speculation and revision as the disease spreads.
- There is a high fatality rate among those with 'full-blown' AIDS.
- Young and otherwise healthy adults are the targets for the disease.
- Infection has a 'sensational' connotation attached to it, and the spread of infection has erroneously been associated with casual non-sexual contact, such as drinking from a chalice cup or swimming in a public bath.
- Infection has been associated with aspects of sexually transmitted agents (bodily fluids), leading to an inherent confusion in identifying risk factors.

An additional characteristic of the disease is the large number of people who are not infected, yet who fear infection because they have the potential to contract the disease. Many gay people fall into this group of 'worried well', since the disease is spread through bodily fluids, notably semen. Faulstich warns within the context of the gay collective that 'although some reactions of the "worried well" may be within normal limits, some of these individuals suffer from sufficient distress to severely disrupt social and occupational functioning' (Faulstich, 1987:552).

Martin and Vance, writing from the perspective of life stress and illness, believe that the characteristics of the AIDS profile

> appear to form a stress-inducing configuration of sufficient seriousness to: (a) cause increased rates of psychological and emotional distress, and (b) lead to changes in various aspects of lifestyle, most notably sexual behaviour, although other areas of individual and social functioning may be affected as well (Martin and Vance, 1984:1 306).

The overriding uncertainty about the AIDS issue in effect lies in the subjective responses of individuals or communities to the issue of *risk*. Risk factors might be objectively identified, but the negotiation of such factors (such as more than one sexual partner, or being in intimate contact with an infected person), could well be overemphasized or underestimated by some people. The subjective process of interpreting risk factors becomes a plausible psychological defence to ward off the anxiety of the AIDS scare. Joseph and his associates give credence to this point. They emphasize that, within the gay community, the nature of risk is often dealt with as 'a rephrasing of a longstanding homophobic viewpoint that homosexuals engage in dirty and dangerous behaviour' (Joseph et al., 1984:1 300). Risk is negotiated by externalizing the threat. For instance, some homosexuals project blame on to societal homophobia, instead of realistically appraising the facts themselves.

Attached to the awareness of risk is a sense of universal betrayal. Heterosexuals feel that their sexual boundaries have been infiltrated, since the generic risk factor suggests that any deviation from monogamous sexual behaviour might be dangerous.[3] Any sexual behaviour outside a long-term monogamous relationship must be accompanied by precautionary measures. Homosexuals feel betrayed because their sexuality has been exposed and analysed according to strict medical principles and moral overtones. Because homosexual sex behaviour has been exposed, what was once a 'closed book' has become public knowledge, with resultant fear, panic, and disgust reflected in responses from scientists and the public alike.

Paradoxically, homosexuals are being told 'how to do it' by people (medical scientists) who have hitherto pronounced homosexual sexual proclivities to be unsound. Distress, based on risk evaluation, can lead to extreme reactions by people who have to alter longstanding patterns of sexual activity that have become an integral part of their lifestyle, and to modify patterns of expression. This is experienced as a change in their identity. The homosexual population being recognized (in the Western world) as the primary target for AIDS, counteracts in three ways:

▶ by responding to world opinion as if it were an onslaught against homosexuality;

▶ by denying the whole issue, and continuing a lifestyle familiar to them before the advent of AIDS; and

▶ by incorporating (ingesting) the symbolic fear into the self, for example the medical uncertainties lead homosexuals to monitor their sexuality, and to view the general lifestyle of the sub-culture in an extremely critical manner. Such behaviours could include celibacy; attacks on homosexual lifestyles (intra-culture phobia), which causes self-imposed estrangement behaviour; and overcompensating for a lack of ongoing intimacy with others by self-obsessive behaviour. Examples include panic attacks, generalized anxiety, hypochondriasis, excessive somatic preoccupation and fear of disease, and attempts at or thoughts of suicide[4] (Buckingham and Van Gorp, 1988; Faulstich, 1987; Flavin et al., 1986; Harrowski, 1987; Holland and Tross, 1987; Lopez and Getzel, 1984, 1987; Pleck et al., 1988; Quadland and Shattls, 1987).

The following case extract illustrates fear of AIDS and attempted suicide:

> A thirty-two-year-old man made contact with one of the writers after a serious suicide attempt. The precipitant to the suicide was fear of AIDS. The client had no symptoms; nor had he been tested for the presence of positive antibodies. The dynamics underlying his suicide attempt included the following:
>
> ▶ recent experience of surfacing fantasies of a homosexual nature;
>
> ▶ fear of verifying these persistent fantasies, which led to a withdrawn, speculative state of mind;
>
> ▶ external issues, such as the morality of homosexuality; fear of being found out at work as well as parental-family censorship was strong; and
>
> ▶ panic at being exposed to the homosexual sub-culture caused a sense of immobilization and self-disgust.
>
> On the eve of his suicide bid, he was confronted by a group of friends, including two homosexual men, at a dinner party. One of the men leaned over and kissed him (frivolously). His immediate response was to leave the room and vomit. He subsequently left the party and went home.
>
> The fear of an 'intimate' expression, as represented by the kiss, reinforced his anxiety pertaining to his own homosexual fantasies, and was transposed on to AIDS.

In the decade since it was first reported, there has been uncertainty, at both the political and the personal levels, about whether AIDS is a gay issue or a public health issue. Because of this, the issue of risk has become confused. AIDS has been universalized as a 'homosexual plague', and homosexuality has been identified as a risk phenomenon. Together, these two features have obscured the full implications of the transmission of the illness. Homosexual lifestyles, which include varying degrees of sexual intimacy, are not necessarily the primary cause of AIDS (McKusick et al., 1985b). The risk factor revolves around *sex behaviour*. The labelling of homosexuality as sex behaviour *only* has reintroduced confusion and stigma in both heterosexual and homosexual thinking. Indeed, from an international perspective, Altman believes that the link between homosexuality and AIDS has the potential for 'unleashing panic and persecution in almost every society' (Altman, 1986:187).

Weeks deals with this issue in the context of generalized societal flux, particularly in the framework of fear, panic, and homophobia. He writes:

> The mechanisms of a moral panic are well known: the definition of a threat in a particular event (a youthful 'riot', a 'sexual scandal'); the stereotyping of the main characters in the mass media as a particular species of monsters (the prostitute as a 'fallen woman', the paedophile as a 'child molester'); a spiralling escalation of the perceived threat, leading to the taking up of absolutist positions and the manning of moral barricades; the emergence of an imaginary solution — in tougher laws, moral isolation, a symbolic court action; followed by the subsidence of anxiety, with its victims left to endure the new social prescriptions, a social climate or legal penalties. In sexual matters, the effects of such a flurry can be devastating, especially when it touches, as it does in the case of homosexuality, on public fears, and on an unfinished revolution in the gay world itself (Weeks, 1985:45).

Weeks's powerful statement reflects the meta-crisis of AIDS. As long as AIDS remains closely bonded to homosexuality, as opposed to sex behaviour in general, those who deny or distort the validity of homosexuality have access to powerful ammunition.

The linking of AIDS to a minority group has far-reaching ramifications for the psychological homeostasis of the group. Again the giant paradox surfaces, to reveal a collective of people struggling with identity issues now also being confronted with an awesome fact: AIDS is primarily transmitted through sexual intimacy, yet sexual intimacy has been regarded as the panacea for consolidating sexual identity. Deductive reasoning therefore undermines previous attempts to promote the reality that homosexuality is no more and no less than a sexual proclivity between men, and reinforces the notion that the homosexual is a carrier of a 'venereal disease' of fatal proportions. Labelling theory can be used to develop a deeper insight into the foregoing contention:

▶ Using Goffman's (1963) view of 'spoiled identity', it can be postulated that homosexuals are, once again, assimilating society's view of stigma, before they identify themselves as so stigmatized. The incorporation of stigma associated with AIDS leads to a disapproval of self, and of others like the self.

▶ Attribution theory as cited by Karr suggests that 'the perception and evaluation of an individual are in part a function of the personality of the perceiver and the social situation in which the perception takes place' (Karr, 1981:3). In respect of AIDS, this reflects on the parameters of 'social distance' and 'social intimacy'. Both are experienced from within and without the homosexual framework. Heterosexual perceptions of homosexuality are distorted with images of contagion and blame, and homosexual perceptions of homosexuality are distorted with images of generalized homosexual homophobia (including contagion). Specifically, if homosexuality is measured in terms of its sub-culture and its fringes, then that culture will, too, become a victim of stigma and oppression. AIDS has moved from an individual-related issue to a collective-related issue, influencing the levels of distance and intimacy. The direction and extent of homophobia, a social distance phenomenon, are affected by sexual suspicion, fear of so-called promiscuity, rumours attributing symptoms to people (e.g. weight loss), and blame.

Self-esteem, an important feature in identity development as discussed in Chapter 1, can be a major casualty of the AIDS scare. Shame, guilt, self-reproach, and embarrassment often occur as a result of societal predisposition to blame the victim (Anderson, 1982:152). Threat to self-esteem is directly experienced. The view of AIDS as a predominantly collective gay disease incurs a typical victim-type response when AIDS is perceived by homosexual men as a threat to their sexuality or sexual identity formation. AIDS may produce ongoing psychological battering because the victim response is, among other things, dependent on the appropriateness of other people's reactions (Sutherland and Scherl, 1970).

According to Anderson, assault trauma, in the context of either physical or psychological assault, can increase psychological distance between a gay man's public and private self, adding to the distress surrounding identity issues. In addition, feelings of rage towards AIDS assailants, in this case directed towards (a) the virus, (b) the homosexual collective (carriers or potential carriers of the virus), and (c) towards society for condoning or encouraging the sense of victimization, lead to depression, chronic low self-esteem, and acting out behaviour. Anderson links the chronic aspect of self-esteem loss to 'manifestations of underlying anger' (Anderson, 1982:153). However, difficulty is often experienced in the expression of anger and rage because AIDS sufferers (including the HIV-infected) and those who represent the 'worried well' may be particularly threatened by other gay men, and, according to Anderson, will often disavow their sexuality and their right to enjoy sex, devaluing their own needs in the process and thus maintaining low self-esteem profiles.

The implications of the above for resolving crisis and consolidating identity are paramount, and are captured in the following statement: *the behavioural and psychological responses to AIDS can also be regarded as a conscious defence against patterns of intimacy.* Coming out problems, such as negotiating the sexual fantasy structure and dealing with the sub-culture, are avoided. What was recently regarded as acceptable within the boundaries of the homosexual sub-culture is viewed with a sense of suspicion because of AIDS, and people are reluctant to deal with aspects of their homosexuality because of the sexual implications of the disease. The need to consolidate identity issues (as discussed in Chapter 1) has been negated or distorted by widespread paranoia. Conversely, overindulgent sexual behaviour occurs as well. This defiant pattern of behaviour persists in those who deny the seriousness of the facts about the transmission of the illness. The following vignette illustrates this in terms of the South African situation:

> A steam bath, located in the centre of Johannesburg and catering exclusively for the gay collective, operates daily. Upon admission (which is by membership recommendation only) each customer is supplied with a towel, a clothes-locker key, and a free condom. Issuing the contraceptive is an attempt to prevent unsafe sex practices, but it is regarded by many as 'an invitation to participate in anal sex'.
>
> The architecture of the steam bath is such that private cubicles afford patrons the opportunity of selecting a partner for sexual interaction in relative privacy. There are community rooms (described as TV or relaxation rooms) as well, where groups of people can participate in group sex. Of note are the number of people who participate in mutual receptive anal intercourse without the use of the condom.[5] Either the serious implications of safer sex practices have not reached the South African population, or disregard of the situation is prevalent, or both.

Characteristics of the AIDS crisis within the homosexual context

An overall characteristic of the AIDS crisis is that medical science initially identified the illness primarily in gay men. Aetiological factors suggest that transmission is dependent on the type, nature, and frequency of sexual intimacy. Recipient anal intercourse is the primary danger. Sub-clinical infections, exposure to sexually transmitted diseases (such as herpes, syphilis, and gonorrhoea), as well as a multiplicity of sexual partners give rise to clinical concern as well. These facts, exposed by the medical fraternity, must be related to the idiosyncratic features of the gay collective. Joseph and his colleagues (1984) identify the fundamental importance of moral and medical attacks on homosexuality from a historical perspective. Thus it is not surprising that the onset of the AIDS crisis was recognized or interpreted by homosexuals as the re-emergence of social homophobia.

Crisis and distress are related (Hoff, 1978). Distress in respect of AIDS is manifested in physiological, behavioural, and psychological signals, such as the following:

▶ Precursors of the AIDS syndrome are common (for example weight loss, fatigue, recurring night sweats, skin rashes, common cold-like symptoms, and diarrhoea).

▶ Anxiety and panic are shown in dealing with the variable aspects of sexual behaviour, since sexual activities have been identified as the main transmission route.

▶ Fear or distaste of safer sex practices, such as the use of a condom, and deliberate attempts to avoid body fluids, induce anxiety about 'spontaneous' sex interplay, which is consequently inhibited.

▶ Difficulty in the negotiation of sex roles. Assumptions about sexual styles and preferences can no longer be taken for granted, and this has led to hesitancy and a reluctance to identify and deal with this issue in a contractual way, particularly when new or unfamiliar contacts are encountered. Moreover, attacks on libertine behaviour, associated with norms of permissiveness, raise fundamental issues about identity, allegiance to the sexual stronghold of the sub-culture, and the persistent desire to maintain links with the sub-culture in order to preserve the continuity of homosexual experiences. The crisis is, in effect, a 'double bind'. Homosexuals are, on the one hand, careful not to criticize publicly those known to have multiple sexual partners. On the other hand, an element of backlash is evident, involving bewilderment, withdrawal from gay experiences, or censure. Confusion, a principal feature of the state of crisis, coexists within the arena of sexual expression.

Crisis intervention and AIDS

It is suggested that the reader peruse the case study of Robert X (Appendix A), before embarking on the ensuing section. In line with the earlier discussion of crisis intervention in Chapter 2, the AIDS crisis is examined below in terms of Golan's five stages. Where relevant, the clinical features of this model will be illustrated by reference to the record of Robert X. Discussion of each stage will also include an overview of some of the main intervention considerations which may guide health care professionals[6] when they are confronted with people who have AIDS concerns.

The hazardous event

A crisis may be anticipated or unanticipated. The hazard, an integral component of crisis interpretation, needs to be identified before negotiating any type of assessment or therapeutic regimen. Where AIDS is concerned, the hazard was initially unanticipated, resulting in a period of shock, denial, and disbelief. The origin of the hazard — identification of the presence of the virus — still forms a threat to the ongoing homeostasis of the homosexual population. The interpretation of the hazard is thus multidimensional, and is concerned with the following:

▶ the negotiation of the process of developing a homosexual identity;

▶ the actualizing of homosexual identity by means of sexual contacts;

▶ a recognition that bodily fluids could contain the virus;

▶ a recognition that unsafe sex practices could jeopardize health;

▶ an inability to determine risk factors; and

▶ a generalized homophobia.

In Robert's case, the generic AIDS scare is evident, including his response to the historical onset of AIDS. The time span includes the present and the past: there is a sense of the immediate, but also pertinent retrospective features. The hazard was experienced in terms of the deluge of reports about AIDS from about 1983 onwards, reports about the first sufferers of AIDS in Cape Town in 1985, and the beginning manifestation of physical symptoms akin to the ARC profile.

Intervention considerations The hazard, alongside the precipitating factor, is the primary clue to crisis identification. Because of the initially unanticipated nature of the AIDS crisis, those infected in the early stages of the crisis responded differently from those who, taking risk factors into consideration, might have been able to prevent infection. Thus the hazard incorporates levels of envy, anger, and resentment towards a sexual lifestyle that was ostensibly 'safe' before the onset of AIDS. A group dichotomy therefore exists between pre-safer-sex sufferers and post-safer-sex people.[7] Therapists should go carefully into the time factor, and determine a cut-off point in defining the 'AIDS era' for those who are experiencing symptoms. Sufferers of the pre-safer-sex era exhibit anger towards the 'lucky' ones, and tend to display a severe 'psychological autopsy' of their previous lifestyle. A history of significant episodes in the individual's past, and the way in which he interprets them, needs to be explored so as to assess any 'unfinished business' related to homosexual issues.

The vulnerable state

All vulnerable moments have a profile of subjective and internal responses. Known as the 'internal dialogue' phase, the individual (or group) weighs up the hazard according to the defence system in operation at the time. Traditionally, the onset of the crisis weakens the defence structure, so that a period of vulnerability and openness to intervention exists. It is clinically important to recognize that the sense of vulnerability vacillates according to internal and external contingencies. In respect of AIDS, the subjective responses to the hazard have particular characteristics, identified by Dilley et al. (1985), Ferrara (1984), Goulden et al. (1984), Joseph et al. (1984), Martin and Vance (1984), Miller (1987), and Millar and Brown (1988), as follows:

▶ uncertainty, with pervasive feelings of anger and anxiety surrounding illness and treatment;

▶ isolation, and fears of social abandonment;

- perception of the illness as a retribution, with overgeneralized feelings of guilt towards homosexuality;
- a difference between recently diagnosed patients and those diagnosed during the first stages of the appearance of AIDS, which may be expressed in a number of areas, such as
  - feelings of anger and betrayal;
  - relentless searching for explanations;
  - feelings of sadness and depression;
  - isolation from families; and
  - the need to undertake life reviews, and an exaggerated sense of unfinished business (Dilley et al., 1985:84).

In examining the vulnerable state, and heeding Golan's (1978) warning that the phases of crisis identification cannot be isolated, dealing with this period must include aspects of the precipitant (the realization of crisis danger), and a recognition that a sense of vulnerability exists in *all* the stages of crisis identification.

The dilemma experienced during this phase by Robert may be likened to a pendulum. The vacillation over risking a test to determine whether or not the antibody to the virus was present (in the face of a drive by the gay community to be tested), and subsequently discovering HIV symptoms, created a long period of indecision. Robert's vulnerability was exacerbated by his lover being diagnosed as sero-positive, with symptoms. In Robert's case, the sense of vulnerability extended over a long period. As Golan (1978) points out, the extent of the vulnerability provides the person with a level of challenge, danger, opportunity, and imminent loss. An element of risk is also involved. However, both Robert and his lover were so harnessed by feelings of fear and danger that there was no likelihood of 'opportunity' or 'challenge'.

Intervention considerations During this phase, all intervention with HIV/AIDS patients requires careful consideration of a range of psychological issues (Stulberg and Smith, 1988). These issues have been systematically identified by Miller (1987), and are listed below.

- **Shock:**
  - about the diagnosis and possible death; and
  - about loss of hope for good news (particularly in respect of a cure).

▶ **Anxiety and fear:**

▷ of the uncertain prognosis and course of the illness (this is specifically related to the complex set of symptoms experienced by AIDS sufferers);
▷ of the prospect of disfigurement and disability;
▷ of the effects of medication and treatment;
▷ of isolation, abandonment, and social/sexual rejection;
▷ of infecting others and being infected by them;
▷ of the lover's ability to cope; and
▷ of loss of cognitive, physical, social, and work abilities.

▶ **Depression:**

▷ over the 'inevitability' of physical decline, and loss of body image;
▷ over the absence of a cure;
▷ over the prospect of the virus controlling future life;
▷ over limits imposed by ill health, and possible social, occupational, emotional, and sexual rejection; and
▷ because of self-blame and recrimination for having been vulnerable to infection in the first place.

▶ **Anger and frustration:**

▷ over inability to overcome the virus;
▷ over new and involuntary health and lifestyle restrictions; and
▷ at being 'caught out' over the uncertainty of the future.

▶ **Guilt:**

▷ about past 'misdemeanours', resulting in 'illness punishment';
▷ about the possibility of having spread the infection to others; and
▷ about being homosexual or a drug user.

▶ **Obsessive disorders:**

▷ relentless searching for new diagnostic evidence and for bodily symptoms;
▷ faddism over health and diet; and
▷ preoccupation with death and decline, and with the avoidance of new infections (Miller, 1987:1 673).

Miller's profile serves to capture the crisis proportions of AIDS. In the vulnerable state, these proportions become amplified, especially where risk, anticipated gain, and opportunity for hope are concerned. Another feature of the vulnerable state is that aspects of personal vulnerability become linked to a sense of collective loss (Holland and Tross, 1987; Rosenbaum and Beebe, 1975; Hoff, 1978). The loss of personal coping ability, coupled with other state or object losses, becomes a 'life predicament'. The life-threatening aspect of AIDS 'leads naturally to the idea that crisis strikes not just a single individual but a unit of interacting individuals' (Rosenbaum and Beebe, 1975:12).

AIDS has been linked to sexual behaviour, and in particular to homosexuality (Kingsley et al., 1987). It is, however, vital to separate sexual behaviour from homosexual experience, which includes cerebral, fantasy, and ideological manifestations of behaviour. The sense that homosexuals have of their vulnerability may be distorted by rumour and sensation, and may have no direct bearing on the individual's present position. Facts must be separated from generalized fantasy. The vulnerable stage in crisis gives immediate access to hidden fears, and accumulated or unresolved life agendas. In fact the vulnerable stage might be the precursor not of HIV infection or AIDS itself, but of some quite different issue, as is shown in the following case vignette from the files of one of the writers.

> On the recommendation of a physician, a couple who both registered HIV sero-positive sought counselling. Both men were in their mid-thirties, and had been together as lovers for over three years. Two sessions were used to examine their emotional fears and anxieties about AIDS. The safe therapeutic situation provided relief for both in being able to express their emotions. When the issue of intimacy arose, however, particularly in respect of crying behaviour and responding in a caring fashion, both men acknowledged that they were unable to cry openly in the company of the other. When the therapist opened up a new avenue of intimate expression, in the context of sadness, the sense of AIDS vulnerability was replaced with a new set of vulnerable feelings. Intimate patterns of expression that had hitherto been taken for granted in the relationship now appeared renegotiable. Upon further exploration, it emerged that both had stereotypical fantasies about how a 'man' ought to behave.

Vulnerability induced by AIDS can transcend the immediate fears of debilitation into the wider arena of interpersonal relationships.

The hidden 'opportunity' factor (and the potential of gain) involved in the vulnerable state is overshadowed by persistent obsession with the illness. Therefore, in treatment, the need to defuse the recurring tension about AIDS must be accompanied by the need to deal with intimate factors confronting the individuals in the context of their socio-ecosystem. The case vignette above also shows that vulnerability is seldom caused by a single factor on its own. The clinician may have to identify a number of features that contribute to the ongoing sense of vulnerability.

The precipitating factor

Besides being the most important diagnostic clue to understanding the crisis, the precipitant (which induces recognition of a state of crisis) normally motivates the person to seek help. The precipitant identifies the 'lowest ebb' period, and forms part of the challenge or risk profile discussed in Chapter 2.

An AIDS crisis might be precipitated by a single hazard or a collection of hazards. These might include the onset of symptoms, a positive response to HIV antibody testing (Helquist, 1987), legislation dealing with infectious diseases, a newspaper report, the loss of a lover, or witnessing others who are afflicted with the illness.

In respect of the collection of hazards, the precipitant may unleash a series of collected behavioural and psychological responses. Part of this phenomenon has been described by Rosenbaum and Beebe as a response to an ecostrain. The precipitant might be induced by a ripple effect in a long wave of strain that has been triggered by a major shift in the functioning of a group or system (Rosenbaum and Beebe, 1975:38). Hence the dispensing of safer sex kits to clubs and bars by the local chapter of GASA in Cape Town could precipitate a response in an individual (or group) to the seriousness of the broader parameters of the AIDS crisis. 'Crisis is news of a difference affecting the patterns of relationships within an ecological group' say Rosenbaum and Beebe (1975:14). Therefore the diagnostic and treatment implications of the precipitant must be dealt with within Hollis's context of 'the person, the problem, and the situation' (Hollis and Woods, 1983). Like the hazard and the vulnerable stage, the precipitating factor(s) can be linked to previous episodes, and not only to the issue that has enabled the person (or group) to identify the source of immediate pain.

In Robert's case, the precipitating factors include the ARC symptoms experienced by both him and his lover Peter, as well as the onset of AIDS in Peter culminating in hospitalization and eventual death. A further precipitant which led to Robert's crisis was the fact that he was away when Peter died. Thus the precipitant was not confined only to the advent of AIDS, but included a circumstantial (i.e. external) event as well.

Intervention considerations A major treatment objective is to allow the person to re-experience the peak of tension in the crisis. Equated to Zimbler and Baring's (1975) critical stage theory, the precipitant gives access to immediate problems, and helps identify specific parameters of stress. Expectations about outcome should be avoided during this period, for they can detract from the immediacy of the situation. The person must be given the opportunity to ventilate freely within the boundaries of controlled catharsis (Rosenbaum and Beebe, 1975; Kadushin, 1983).

This period facilitates the person's sense of urgency by helping to partialize, focus, and capture the salient features of the crisis. Because of the influence of AIDS on individual people within the homosexual ecosystem, and also the differences in individual personalities, each person will experience the precipitant uniquely. Intervention protocols should therefore involve the assessment of each of the following:

▶ the severity of stress factors, including suicidal ideation, or damage to self and others;
▶ identification of the precipitant with reference to the person's social system;
▶ a clinical overview of the person's current defence profile, with possible access to previous coping strengths;
▶ the person's internal frame of reference *vis-à-vis* his understanding of the dilemma;
▶ the ability to deal with AIDS panic; and
▶ the validity of the precipitant.

In respect of the validity of the precipitant, the presenting issue may not represent the real problem. The presentation of any feature of AIDS as the primary precipitant might in fact mask the serious deliberations of identity development, relationship difficulties, and depression. This particular point has implications for the 'coming out' syndrome. Recent clinical experience has indicated that it is often safer to talk about HIV infection and/or AIDS than to deal with the coming out process. Therapists need to be aware of this deflecting pattern of responses by some clients. Hence the AIDS issue may surface as a conscious defence during this period, and act as a barrier to dealing with other issues.

During this period, the therapist needs to draw attention to parameters and guidelines for safer sex practices, and gently pursue the client's attitude and response to safer sex objectives.

The state of active crisis

This period essentially addresses panic, disease, personal and social loss, and the implications of disfigurement and death. The state of active crisis has definite symptoms, and aspects of psycho-social deterioration are obvious. Anxious intropunitiveness is evident, including self-condemnation, lack of insight, apprehension about the future, and generalized anxiety.

The active crisis is normally a fluctuating process, unlike the set period of time described in much of the writing on crisis intervention. The reason for this clinical idiosyncrasy lies in the differing experiences of immediate loss and accumulated loss. This is explained in terms of the buffering components of AIDS and the intrusion of new sets of hazards and precipitants into the overall crisis scenario. For example, an individual might be in crisis because of the death of a lover. A new hazard intrudes upon this individual when he discovers that he is HIV-positive. His response (vulnerable state) to this fact is compounded by the death of his lover, coupled with guilt and the fear of contagion in future relationships. Active crisis is never separate from the previous stages of crisis identification, and is inevitably influenced by emerging internal and external stress factors. An apt description of this period is 'bereavement overload', and according to Holland and Tross it is fraught with recurring suicidal thoughts as a means of possibly avoiding the intolerable consequences of the progression of the disease. They urge clinicians to note that the accumulation of loss factors, exacerbated by anticipated loss, leads to expression of suicidal ideation to a far greater extent than in cancer patients (Holland and Tross, 1987).

The death of Peter, and Robert's physical and emotional deterioration, placed Robert in a state of active crisis. This period, according to Zimbler and Barling (1975), includes the critical phase of crisis resolution, which relates particularly to depression or depressed behaviour. Because depression incorporates the lost individual (or object), the anger towards the lost object is turned inwards and results in ingested anger, guilt, and self-reproach. The critical phase in the crisis experience undergone by Robert has a cumulative toll. The loss of a lover, the loss of a relationship, hospitalization, and the loss of employment coupled with the loss of independence are superseded by the recurrence of AIDS fantasies. Thus the anticipation of gain through challenge is lost as well. As Zimbler states:

> Constituting depression as such out of the suppression of feelings, leads to a view of depression not as a feeling in itself, but as the pursuit of a state of non-feeling. In this sense it does not present a picture of the crisis state, but rather it suggests the avoidance thereof. The rather radical implication of this dialectic is that the act of clinical diagnosis of depression facilitates dependency on the part of the client, and forestalls the client's own tendency for self-regulation and his potential for crisis resolution ... thus begins the vicious spiral of descent to deeper levels of depression and more profound helplessness (Zimbler, 1981:780).

The dilemma of Robert's state of active crisis, besides the experiencing of severe depressed feelings (in particular about his relationship with the lost object), is that the crisis is experienced as multifactorial. In accordance with the protocols of risk described in Chapter 2, part of the meta-crisis experienced by Robert was the exposure of his homosexual world (otherwise an area negotiated at his own choice) to a heterosexual world which he perceived as apparently hostile. Hence, information needed by medical staff was refused by him for fear of 'punishment'.

Intervention considerations The overall therapeutic task is to recognize the multifactorial implications of the crisis stage (Slaikeu, 1984). The intervenor will have to deal with crisis defusion at different levels, and in each session to deal with all of the patient's major problems in serialized fashion. By this is meant that the main problems should each receive attention with an equal sense of priority, so that the client does not feel that one or more problems may conveniently be forgotten. This may be likened to the script of a radio or television serial where each character has a role to play in each episode of the serial.

The stage of active crisis is fraught with potential for emotional flooding. This period incorporates past, present, and future material. The intervenor must be able to deal with physical, sexual, self-esteem, and state and object losses, while at the same time reinstating a sense of homeostasis (Nichols, 1985). An active crisis, if it leads to therapy, will in due course result in relief and then hope, thus beginning the last stage of crisis management — reintegration.

The stage of reintegration

This period usually flows from a process of treatment and recovery. In the context of AIDS, it is dealt with through the process of acquiring levels of acceptance and adjustment. The time span linked to this phase reflects both the present and the projected future.

Reintegration, hope, and cure are part of Robert's adjustment phase. However, this period is continuously contaminated with further crises manifesting themselves. Alongside the notion of being able to relinquish the lost object, and so make place for anticipated gain, Robert has thus far been unable to relinquish certain aspects of his lost lover. Clothes, pillows, photographs, and the like constantly re-evoke his period with Peter. In different words, his memory of Peter symbolically perpetuates the state of ongoing crisis. Coupled with this is his constant reflection on how the relationship *ought* to have been. The reintegration phase, paradoxically, is submerged under the fantasy anticipation of correcting a relationship that during its nesting period did not fulfil his expectations. A phantom relationship with a deceased object coincides with the reality of his present predicament. The last stage (reintegration) is beset with external realities which include his receiving a disability grant, food parcels, and the gradual detachment of certain friends. Furthermore, physical debilitation and the confusing desires of hope exist, mingled with a feeling to terminate living. Therefore reintegration in Robert's case is far from complete, and vacillates between relapse into crisis and periods of self acceptance.

Intervention considerations The intervenor, during this phase or period, must be able to distinguish clearly between the complexity of the AIDS-related clinical symptoms and the person's current reality base. As stated previously, the period of reintegration is dependent upon both the present and projected future aspirations of the client.

In the first instance, the intervenor, in order to determine the current reality base, must pay attention to the following variables:

▶ Has the diagnosis been clinically confirmed?

▶ Has the diagnosis changed? (It is not uncommon for a person to be diagnosed as a 'fully blown' AIDS patient, but subsequently to revert to the ARC period.)

▶ Is the person being treated as a 'fully blown' AIDS patient, an ARC sufferer, an antibody-positive person (with or without symptoms), or as a 'worried well' person?

Secondly, different strategies for the reintegration period are required by the intervenor, which are related to the accurate assessment of the above-mentioned variables.

Clearly, a person living with AIDS will experience reintegration along the lines expounded by Kubler-Ross. Dealing with death and dying is a focal point during the terminal phases of AIDS, and in this regard Deuchar (1984) supports Kubler-Ross's theoretical application to the stages of dying. In addition, he notes that the essential aspects of consistent care, reality-based information, treatment compliance patterns, pain relief, and dealing realistically with the fluctuation and abandonment of hope, must also be handled.

In this respect, the writers note with concern the looseness with which the term AIDS is used by health care workers and the public alike. The ultimate collapse of the immune system as a result of the AIDS virus does not always necessitate the clinical label of 'AIDS'. AIDS does not kill. Opportunistic infections, the like of which include protozoal infections such as pneumocystis carinii, viral infection such as cytomegalo virus, bacterial infections related to tuberculosis, fungal infections like candidiasis, and cancer (Kaposi's Sarcoma), result in death. These infections are however rendered deadly because of the interference of the AIDS virus with the body's own defences.

The implication of the foregoing discussion is that individualized assessment of the previous four stages, based on solid clinical enquiry, is critical to successful reintegration. Also required is direct confirmation of the person's medical and immune status, which need to be checked and then reconfirmed at regular intervals.

The idiosyncratic features of the reintegration phase in the AIDS crisis should be dealt with in terms of the following two principles:

▶ No single clinician may 'own' a patient. Team work is essential. This allows the person to be treated holistically and, as a result, dependency needs can be dealt with. Transference/countertransference features are therefore placed in context.

▶ Reintegration in respect of people with full-blown AIDS has a definite termination emphasis linked to it. Unlike the condition of HIV-antibody positive, or ARC, which has an ongoing and fluctuating 'life span', AIDS, according to Perry and Markowitz (1986), should be strategically dealt with as a terminal disorder.

In conclusion, the writers support Knobel's (1986) emphasis that intervention is incomplete without the attendant fundamentals of education and prevention, which include safer sex campaigns, HIV-antibody testing, and community education. While this book does not directly address education and prevention strategies, the link between intervention and prevention-education must not be underestimated. Ineffective prevention-education leads to increased infection, while ineffectual care strategies jeopardize effective education (Pegge, 1988a).

The crisis of AIDS, both at the micro (individual) and macro (community-society) level, can be dealt with by the application of crisis principles. In particular, an element of danger and threat coexists with an apparent inability to perceive a realistic outcome of hope. In addition, an overwhelming feeling of loss or impending loss is evident. The accrual of such loss deflects from the natural consequences of mourning, and individuals (as well as groups and communities) are liable to become fixated at the loss level. All crises, be they micro or macro, have an identifiable and causative precipitant which, when explored, facilitates access to conscious and unconscious material. In the area of hope, the crisis of AIDS stimulates human endeavours such as networking, relationship building, and self-help groups, which surface during the period of crisis resolution.

Notes

1. This book addresses the impact of AIDS on the homosexual collective only. Nevertheless, the assumptions and principles of crisis and crisis intervention can be applied to any AIDS crisis situation.
2. Health care professionals are cautioned against comparing AIDS with other illnesses. Such comparisons can detract from the immediacy of AIDS, and minimize the consequences of the illness.

3. See, for example, T D Moodie's 1988 study, 'Migrancy and Male Homosexuality on the South African Gold Mines', *Journal of Southern African Studies*, 14(2), 228–56.
4. W S Meyer, in his 1988 work 'On the Mishandling of "Anger" in Psychotherapy', *Clinical Social Work Journal*, 16(4), 406–17, notes that clients suffering the pressure of denied or suppressed anger can be easily stimulated to overburdening levels of rage. Such intense effects are typically accompanied by fear, panic, and ultimately a sense of annihilation.
5. This perturbing feature was reported to one of the writers in an interview with the manager of the steam bath.
6. The terms intervenor, therapist, clinician, and health care worker are used interchangeably by the writers.
7. For details regarding safer sex guidelines, refer to the *AIDS Procedure and Information Manual*, Cape Town: GASA 60-10, 1989, page 13.

6 Development and nature of the formal gay movement in South Africa

The development and nature of a formal gay movement in any society is influenced by the society's particular structure, and public and legal attitudes towards homosexuals and homosexual behaviour. Hence, while the formal gay movement in South Africa is in some respects similar to gay movements elsewhere, in other respects it is different and uniquely 'South African'.

As a background to tracing these similarities and differences, this chapter commences with an overview of gay movements internationally. Thereafter, focus moves to South Africa. The singular nature of South African society is described, specific attention is given to the country's laws relating to homosexual behaviour, and the development and nature of the formal gay movement is discussed and analysed. Finally, consideration is given to contemporary crises which affect the nature and functioning of the formal gay movement, the homosexual sub-culture, and the identity development of individual homosexuals.

An international perspective

The second half of the nineteenth century saw the rise of the first homosexual movement. In Germany, the work of Karl Ulrich and the subsequent foundation of the Scientific Humanitarian Committee in 1887 were the precursors to gay liberation attempts in Holland, Austria, the USA, Soviet Russia, and England (Mieli,

1980). In other countries, although not by means of formal organizations, attempts were made to secure the human rights of homosexuals. For the first time, cultural and political personalities exposed homosexual problems and concerns almost as acts of open defiance. John Addington Symmonds wrote in 1891:

> We maintain that we have the right to exist after the fashion which nature made us. And if we cannot alter your laws, we shall go on breaking them. You may condemn us to infamy, exile and prison — as you formerly burned witches. You may degrade our emotional instincts and drive us into vice and misery. But you will not eradicate inverted sexuality (Symmonds, cited by Fone, 1980:1).

A literary trend emerged that sponsored human as well as sexual rights. The championing of such rights was reflected in numerous published works, including those of Carpenter, Benkert, Havelock Ellis, Symmonds, Ulrich, and Wilde.[1]

Three clear phases are discernable in the history of the homosexual movement. The first phase, incorporating the works of the writers mentioned above, attempted to demonstrate the trans-historical existence and value of homosexuality as a distinct sexual experience, particularly within the framework of literary culture.

The second phase constituted a scientific revival from a sociological perspective. Authors such as Ford and Beach, Kinsey and his associates, and Hooker set out to comment on the values, traditions, and public attitudes covering the homosexual experience.

The third phase, overlapping with the second, but more vocal and linked to activism rather than a sense of the esoteric, was an attempt to

▶ recapture and reassert positive values of homosexuality;

▶ re-examine research;

▶ locate the sources of social oppression (Weeks, 1981:77–8); and

▶ examine fundamental human rights.

However, the early homosexual rights movement experienced severe setbacks in the 1930s. The onslaught of Nazism, Stalinism, Fascism, and later the ravages of World War 2 wiped out virtually all traces of the first wave of gay liberation. What was left in the wake was a collection of penal codes that, under the umbrella of 'social perils', secured a stranglehold over homosexuals. Post-war legislation in most countries, with the exception of Holland and Japan (Mieli, 1980), was notoriously forbidding. In Western countries, penalties for homosexual law infringements ranged from one year minimum to life imprisonment. In the last decade or so, however, there has been a general relaxing of legal tensions, although the current legal position of homosexuals continues to differ from one Western country to another.[2]

In the United States of America, where legislation differs from state to state, the introduction of several pro-homosexual statutes has been achieved by the National Gay Task Force (NGTF), which is the largest and most well-established of three major homosexual movements in that country (Reuda, 1982). Perhaps the most noteworthy example of successful advocacy is the pressure the NGTF exerted over the American Psychiatric Association to remove homosexuality from its list of mental illnesses. The NGTF is one of the major affiliates of the International Gay Association (IGA), founded in Coventry, England, during 1978. Currently 21 nations are represented in IGA, which includes as 'associate' members 15 other countries including South Africa. According to Rueda (1982), the IGA's activities are directed towards co-operation with the World Council of Churches (support for homosexual rights); Amnesty International (status for homosexuals in jail as 'prisoners of conscience'), the World Health Organization (the deletion of homosexuality from its list of 'diseases and mental defects'), and other international bodies (Rueda, 1982:155–6). It is noteworthy that the IGA has not granted full membership to the Gay Association of South Africa (GASA) because of the worldwide boycott and sanctions against the South African apartheid regime, as well as the fact that GASA is not perceived as an anti-apartheid organization.

The emergence of gay people as a force in the USA, parts of Europe, and Britain is based on gay political activity. Writers such as the Gay Left Collective (1980), Gearhart (1981), Jay and Young (1972, 1978), Marotta (1981), Park (1981), Richmond and Noguera (1979), and Weeks (1977) have examined homosexual politics from a variety of perspectives. The term 'homosexual politics' is an umbrella term, embracing the following:

▶ gay liberation movements;
▶ homophile organizations promoting homosexual needs;
▶ lesbian feminist movements;
▶ radical gay activism;

▶ liberation politics, including civil rights; and

▶ gay task forces.

These are the legacy of pioneering political reformers who maintained that 'calculated political manoeuvering was needed to pass gay rights legislation' (Marotta, 1981:224). The tide of gay liberation or political activism, under whatever operational premise, has one particular mandate: to decriminalize homosexuality and to remove the 'half-caste' status that has beleaguered homosexuals for centuries. From the beginning, the pursuit of gay rights was meant to persuade homosexuals that they had a right to their feelings, and that they could band together to make serious political moves when that right was threatened.

It should be pointed out that although many countries have legalized homosexual behaviour between consenting adults, research reveals that a diverse range of homophobic attitudes is still prevalent. Some years ago a major study by Weinberg and Williams, which examined Denmark, Holland, and the United States, reported an overall lack of public acceptance despite favourable official attitudes to homosexuality (Weinberg and Williams, 1974:86). A more recent study conducted in South Africa under the auspices of the Human Sciences Research Council revealed an overwhelming homophobic attitude in white respondents. Over 70% thought that homosexuality between consenting adults should not be legalized (Glanz, 1987:1).

The history of gay liberation, the formation of international gay groups, and the advent of the second wave of homosexual liberation in the United States, have been described both in popular journals and in a multitude of scientific writings (e.g. Bullouch, 1979; Denneny et al. 1984; Greenberg and Bystryn, 1984; Humphreys, 1972; Jay and Young, 1978; Katz, 1976; Rueda, 1982). Such publications have as their impetus the second wave of liberation attempts, born out of the celebrated Stonewall riots in New York in 1969. Bell captures the spirit of this period as follows: 'The sixties were conception years, prelude to the seventies. As such they were necessary. As a free gay man, I was really born in 1970' (Bell, 1984:29). Bell is referring to the *symbolic coming out status* experienced by gay people. It is this very notion of 'legal freedom' which enables a homosexual identity to be consolidated. This feature is notably absent in South Africa, and thus the formation of a gay identity, as dealt with in preceding chapters of this book, is never a completed process. It continues to be hampered by the absence of legal rights.

The aftermaths of the Stonewall riots, followed later by liberation efforts in Australia, Britain, and Europe, have been documented in detail elsewhere.[3]

Features affecting gay 'liberation' in South Africa

Humphreys, in the context of the debate about gay emancipation, has identified societal means to inhibit homosexuality, and some of their consequences for homosexual people:

> Social oppression, at least as directed against those who reveal a preference for their own sex, takes three basic forms: *legal-physical*, in which certain behaviour common to the stigmatised group is proscribed under threat of physical abuse or containment; *occupational-financial*, limiting the options for employment and financial gain for those so stigmatised; and *ego-destructive*, by which the individual is made to feel morally inferior, self-hatred is encouraged, and a sense of valid identity is inhibited (Humphreys, 1972:9).

Before evaluating the various attempts to initiate, develop, and maintain a semblance of gay liberation in South Africa, a brief overview is given of the main features which determine the mosaic of South African society[4], in an attempt to locate homosexuals within this structure. Thereafter, the legal position of gay people will be discussed.

South African society

Kotze (1975) argues that any gay organization must be influenced by the structure of the parent society. Diversity within the gay subculture reflects a fragmentation similar to that within the parent culture.

Political and social segregation of the races within South Africa existed for many decades before the National Party came into power in 1948. However, the inauguration of 'apartheid' entrenched the politically dominant position of whites, and consolidated the lesser position of disenfranchised Africans, 'coloureds', and Indians within a stratified and racially segregated society. This reality, permeating all aspects of South African life, emphasizes the racial and cultural differences between people. Thus McKendrick specifies the divisive structure of the population (shaped by political ideology), by noting that within South Africa there are 13 artificially created nations: the whites, 'coloureds', Indians, and 10 different African nations, if the so-called independent and self-governing national states[5] are included (McKendrick, 1987b:20).

South Africa, including the homelands and self-governing states, has a population of about 34 million. Its internal policies, underpinned by statute and reinforced by racial prejudice, are based on the classification of its population on a racial basis. When the National Party gained power by a small majority in 1948, the word 'apartheid' (separate development) was born, and this ideology continues to be reflected in the *de facto* domestic policies of the present South African government (although these policies are currently the subject of negotiation and possible revision). Apartheid can be defined, according to Savage, as

> a system of minority domination over statutorily defined colour groups on a territorial, social and economic basis. It embodies structured inequalities which support and favour particular classes and groups of people at the expense of others (Savage, 1986:3).

In the context of apartheid, four major racial groups are officially recognized. This is based on ethnic differences, on the assumption that people of different 'ethnic groups' form communities that 'possess permanent elements of "culture" that cannot and should not be eroded' (Lowe, 1988:21). These four racial groups, and their populations (including the homelands) are:

Asians (about 1 million people);
Blacks (about 25 million people);
'Coloureds' (about 3 million people); and
Whites (about 5 million people) (Eberstadt, quoted by Wilson and Ramphele, 1989:29).

Apartheid legislation, including legislation on residential segregation, influx control, separate amenities, job reservation, and the like, was enacted at different times since 1948, along with security laws empowering the government to keep under surveillance, prosecute, or detain without trial many people opposed to its rule.

In an attempt to redress the disenfranchised position of some people of colour, with particular reference to their right to vote for a government of their choice, the Republic of South Africa Constitution Act, 1983 created a tricameral parliament with three houses: one for whites (House of Assembly); one for 'coloureds'

(House of Representatives); and one for Indians (House of Delegates). McKendrick draws attention to the differential notion of government apropos legislation relating to 'common affairs' (such as the budget on defence matters), and 'own affairs', which concerns matters specifically affecting a particular population group in relation to the maintenance of its identity, and the upholding and furtherance of its way of life, culture, traditions, and customs (McKendrick, 1987b:20–1).

A striking feature of this new dispensation has been its failure to offer blacks any meaningful political role. Of particular importance are the discrepancies, in respect of racial divisions, in welfare, education, health care provision, and housing. This constitutional anomaly has been addressed by Savage, who spells out the iniquitous costs of apartheid, especially the human cost, which is to be viewed and experienced within the daily fabric of South African life (Savage, 1986:4).

A class division based on race has been artificially created by statute. Within a culturally heterogeneous society, various groups of people coexist with their own mores and customs and diverse backgrounds: indigenous Africans, with a rich history of tribal affiliations; intercontinental whites, with a background of Anglo-French-Dutch heritage; Indians, who originally came to South Africa on indentured labour plans; and 'coloureds', who have a variant of the 'white' culture. All are composed of distinct ethnic, traditional, language, and religious affiliations, and each may be divided into smaller parochial groupings. In addition, immigrants in large numbers, including Portuguese, European Jews, Anglo-Saxons, and Greeks add to the cultural diversity of the South African population. The interaction between cultural, social, and political groups in South Africa has to be viewed against the backdrop of certain minority or majority group affiliations. In this regard, Afrikaner white identity is of paramount importance. Based on a strong tradition of fundamentalist religious principles, the content of Afrikaner Nationalism has been summarized as containing three main elements: sacred history, civil theology, and civil ritual (Moodie, cited by Lowe, 1988:37). The most powerful Afrikaans church, the Nederduits-Gereformeerde Kerk (or NGK), has a following of nearly 40% of the total white population (Buis, 1979:102). Of relevance to this group is that the dogmatic aspects of the Dutch Reformed Church's teaching have conveyed certain attitudes towards Africans and other people of colour. This attitude differs radically from the interpretations of other Christians, and expounds that 'Ethnic diversity is in agreement with

God's will' (Buis, 1979:105). Villa-Vicencio refers to this as a 'prescribed state religion', and qualifies his remark by saying that 'Afrikaners applied themselves with a sense of urgency to build a nation with divine purpose and mission' (Villa-Vicencio, 1988:139). It is also appropriate to note that the official policy of the NGK towards homosexuality is that it is a disease 'byna patologiese vrees vir die teenoorgestelde geslag', and that it is the duty of the Church to 'cure' this 'obsessional' behaviour (Exit, 17, Feb./March 1987:2).

It must be remembered that South Africans are certainly divided by denominational structures, traditions, and beliefs. The English-language churches, including the Methodists, Anglicans, and Catholics, have in the past not been guiltless of expounding religious principles in the context of their social, political, and economic interests (De Gruchy, 1985:91). However, they have also been associated with a liberal tradition. In addition to Christianity, two other world religions coexist in South Africa, namely Islam and Judaism, each with a unique interpretation of the social and political issues confronting their congregations.[6] The rise of the African independent churches, such as the Church of Zion, has created another variation in the cultural and religious diversities of South African society. These churches are historically rooted in African pre-capitalist social formations, and their members still cling to some African traditions, most notably ancestor worship, and tribal affiliations (Mosala, 1985:110).

Finally, this brief sketch is not complete without mention of the economic inequalities that persist in South Africa. Historically, the majority of Africans have been trapped in rural underdevelopment or the migrant labour system; more recently increasing numbers of them have been making the painful transition to urbanization. Whites, on the other hand, enjoy comparative affluence; even the poorest of them have been protected by state-created employment. The differential standard of living within the various statutory population groups is illustrated by the fact that the per capita disposable income of whites is well over double that of Indians or 'coloureds', and nearly four times as much as urban Africans (Wilson and Ramphele, 1989:20).

The homosexual collective in South Africa has no monolithic status with which to counteract the fragmentary nature of society. Apartheid ideology has thus enforced separate identities among homosexuals as much as among heterosexuals. All have been victims of the apartheid ideology, about which Posel has commented:

Political, economic, social, cultural, and sexual segregation were cast as divinely ordained, historically vindicated and the foundation of a just and harmonious society. 'Being white' meant being socially and culturally distinct, politically and economically privileged, and physically separated from those who were not (Posel, quoted in Sharp, 1988:81–2).

The legal position of gays

The foregoing sketch of South African society serves as a backdrop against which to examine the legal position of gays in South Africa. At the first National Gay Convention, held in Johannesburg during May 1985, a prominent advocate presented a paper on the legal regulation of homosexual behaviour. He said:

> As a lawyer, it astounds me to read in the standard English-language textbook on [South African] criminal law (B & H, Vol II, 2nd ed., 1982:267) that the crime in South African law of committing an 'unnatural sexual offence' is constituted by any 'gratification of sexual lust in a manner contrary to the order of nature'. The same textbook (pp 270–271), written in 1982, sums up South African public opinion in this area by stating that it is 'not yet ready ... to accept the abolition of sodomy (and other "unnatural" acts) as criminal [even] when practiced in private between consenting adults' (Cameron, 1985:1–2).

Common law offences

South African criminal law consists of common law and statutory law. The common law is based on past legal practice, on the contributions of Dutch and French writers of the seventeenth and eighteenth centuries, and, in the final instance, on Roman law (Joubert, 1985). The historical precedents to the present statutes, which date from 1886 (Cape) and 1898 (Natal), in effect contain similar clauses to present sections of the Criminal Procedure Act of 1979, as amended (Joubert, 1985). One of the provisions which still exists is for the imposition of a whipping over and above fines and imprisonment for gross indecency between two men (Section 293).

'Crimes against morality' can be said to be common law crimes. Two sections of the common law are relevant (Hunt, 1982:271–8):

▶ **Sodomy** Sodomy consists in unlawful and intentional sexual relations *per anum* between two human males. Ejaculation is not necessary, but there must be penetration of the anus. Without penetration, the crime might be that of attempted sodomy. Both the inserter and the insertee are guilty. A boy under the age of 14 years is, however, legally incapable of being the inserter.

▶ **Unnatural offences** An unnatural offence consists of the unlawful and intentional commission of an unnatural sexual act by one person with another person or animal. Coercion might remove unlawfulness from the act. If a 'natural' sexual act was intended, an offence would not have been committed. Some examples of what is and what is not 'unnatural' in South African law follow. One male kissing another in circumstances showing lust is *probably* not an offence. One male touching the organs of another male has been held in court not to be an offence. Mutual masturbation between two males, masturbation of one male by another male, friction of the penis between another man's thighs (intracrural intercourse) or against some part of another male's body, have all been held in court to be unnatural offences. Fellatio (oral sex, also known as 'oral masturbation') between two males may be criminal, and has been held in a court of law to be an unnatural offence.

The requirements for conviction are as follows:

Sodomy:

▶ unlawfulness;

▶ intention;

▶ sexual relations *per anum*; and

▶ between two men.

Unnatural offences:

▶ unlawfulness (in this respect, unlawfulness would be what, in the opinion of the court, runs counter to what public policy would regard as a 'natural' sexual act, and what is in accordance with the law);

▶ intention; and

▶ proof beyond reasonable doubt that the act in question had in fact been committed.

In addition to the above common law offences, there are a number of legislative provisions which are relevant to homosexual sex behaviour.

Statutory offences

The principal legislation concerned is the Immorality Act, 23 of 1957 as amended[7], which stipulates four main offences related, or potentially related, to homosexual behaviour:

▶ Section 14 provides that any male person who commits or attempts to commit an immoral or indecent act with a boy under the age of 19 years shall be guilty of an offence. The maximum fine is R1 000, and/or six years' imprisonment, but if both parties are under 19 years there is no offence.

▶ Section 10 provides that any person who entices, solicits, or importunes in any public place for immoral purposes, or who openly and wilfully exhibits himself or herself in an indecent dress or manner at any place visible to the public, or to which the public has access, is guilty of an offence. The maximum fine is R400, and/or imprisonment for two years.

▶ Section 20 provides that any person who in public commits any act of indecency with another person, or in any way assists in bringing about the commission by any person of any act of indecency with another person shall be guilty of an offence. The maximum fine is R400, and/or imprisonment for two years.

▶ Section 20(A) provides that acts committed between men at a party and which are calculated to stimulate sexual passion or to give sexual gratification are prohibited.

▷ A male person who commits with another male person at a party any act which is calculated to stimulate sexual passion or to give sexual gratification, shall be guilty of an offence.
▷ For the purposes of the Act a 'party' means any occasion when more than two people are present.
▷ The foregoing provisions of the Act do not derogate from the common law, other provisions of the Act, or a provision of any other law. (The maximum fine is R400, and/or imprisonment for two years.)

While some interpretations of the Immorality Act suggest that homosexuality between consenting adults in private is legal (Helm, 1973), Joubert disagrees, pointing out that such acts are common law offences:

Tot so ver gelees, sou 'n mens die gevolgtrekking kon maak dat die regsposisie in SA t.o.v. homoseksualiteit eintlik baie verlig is en dat die feite seksuele omgang in privaatheid tussen instemmende volwasse mans nie strafbaar is nie. Die angel sit egter in bepaling [20(A) (3)] van die Wet: 'die bepalings van subartikel (1) doen nie afbreek aan die gemenereg, 'n ander bepaling van hierdie Wet of 'n bepaling van enige ander wet nie'. En die gemenereg maak alle sodomie en 'onnatuurlike' seksuele handelinge met ander persone strafbaar. In feite is seksuele verkeer tussen twee mans dus in geen omstandighede in Suid Afrika wettig nie (Joubert, 1985:40).

The legal paradox that exists thus lies in the *interpretation* of the law. However, certain clarification is offered by Judges Vermooten and Schabort in an appeal case heard in the Witwatersrand Local Division of the Supreme Court (S v C, *SA Law Reports*, 1987, (2) 76). The judges overruled a magistrate's findings that the appellant was guilty on a charge of engaging in a sexual act (with another person) in a Johannesburg steam bath in the presence of others (police officers).

Part of the summary of the judgement reads as follows:

Section 20A of the Immorality Act 23 of 1957 was not designed to prohibit the acts therein contemplated if performed in private. To warrant a conviction under s20A there must be physical presence of a person or people conscious of the conduct envisaged in the section. *Mens rea* is an element of the offence on the part of the person indulging in the conduct in question. It would seem that the intention of the legislature, according to the nature, purpose and scope of s20A, is that people indulging in the conduct therein contemplated are required to do so with due foresight and care as not to impose their behaviour upon others and as not to expose others thereto. They are in a position to determine the time and venue for their intimate actions and they can ensure that they will take place in private. Accordingly, *culpa* as to the presence of others would provide *mens rea* for the purposes of s20A.

The precedent in this particular judgement needs to be addressed. Firstly, according to the judges, the intention of the relevant section in the criminal law is not designed to prohibit sexual acts between consenting adults in private. A further consideration would be that the appellant and partner admitted to persisting with sex-directed conduct in private *before* the third person entered. The appellant was therefore not charged with committing a crime based on same-gender behaviour (for the reason that a 'party', as defined in the Act, never came about). Finally judgement overruled the contention that the legislature's intention in introducing s20A was to stamp out homosexual gatherings. It was intended to prevent the obtrusion of conduct which, from time immemorial, has to many people been profoundly repulsive as depraved and repugnant to nature. The judgement continues:

> The fact that the private commission of acts envisaged by s20A(1) does not fall within the prohibition, provides a strong indication, in my view, that the presence of people was intended to have a mental element. The commission of acts of this kind in the physical presence of people who are asleep or for some other reason not aware of them (e.g. owing to darkness or blindness or deafness) occurs for all intents and purposes affecting the latter, in private (Vermooten and Schabort, 1987: S v C, *SA Law Reports* (2) 76).

Application of laws

The dilemma which arises from differing criminal and common law interpretations leads to the legal crisis experienced by homosexuals in South Africa. While homosexual acts in private are, in effect, of no legal consequence, the behaviours associated with homosexual practices, such as sodomy, 'unnatural' sexual acts including masturbation, and acts designed to promote 'homosexual behaviour' are proscribed. It is thus legal to be labelled as a homosexual, but illegal to engage in homosexual sex practices. This dilemma is heightened by the number of prosecutions and convictions for such offences. The extent of prosecutions and convictions for sodomy and indecent assault by man on man in the decade 1971 to 1980 is reflected in Table 6.1.

Indecent assaults include statutory rape, exhibition in public, and police entrapment or *agent provocateur* activities.

The fact that aspects of homosexual conduct are proscribed by the common and criminal law, and that a diversity of convictions has been upheld on appeal, creates a general *climate* in which being gay is considered, among other things, as 'unnatural'.[8] Of more serious consequence is the fact that an aura of criminality is attached to the person and his behaviour. This point underpins the pervading sense of crisis (threat and danger) for homosexual people, and contributes to part of the internal process of self-oppression and the constant lack of self-esteem which they experience. Even though a gay person has ostensibly accepted his identity (i.e. identity consolidation), the pervading threat of legal process hampers his ability to become a 'public' homosexual. This point is acknowledged by most of the respondents to the empirical study reported upon in the following chapter. It has therefore been assumed that

> if acting as is natural for a gay person can bring the criminal law into operation, then merely being gay must also be criminal ... the scene is thus set for guilt, anxiety, inhibition and fear on the part of gay people (Cameron, 1985:1).

Table 6.1 Prosecutions and convictions for sodomy and indecent assault by man on man, 1971–1980

Year	Prosecutions Sodomy	Prosecutions Indecent assault	Convictions Sodomy	Convictions Indecent assault
1971	337	89	182	54
1972	307	68	158	35
1973	323	82	178	54
1974	327	74	167	34
1975	323	58	228	38
1976	278	47	200	39
1977	266	37	207	28
1978	289	50	186	31
1979	317	103	226	56
1980	304	109	197	79
TOTALS:	3 071	717	1 929	448

Source: Annual Reports of Criminal Offences (Central Statistical Services: Pretoria, 1982).

The writers draw attention to common law and criminal law interpretations, and the fact that a large proportion of prosecutions and convictions have dealt with police entrapment in *public* areas, sexual behaviour in toilets and similar places, sexual interaction with males below the age of 19 years, and police harassment in gay clubs and steam baths.

Specific attention is drawn to this point, for seemingly there are contradictions between the interpretation of the law, the implementation of the law, and the impressions held by homosexuals themselves. There is no doubt that homosexuals in South Africa do not enjoy the legal freedom of expression experienced by their counterparts in some other countries. Table 6.1 indicates numbers of prosecutions and convictions, but this does not imply that homosexuals are necessarily harassed by legal sanctions. Men and women have enjoyed homosexual liaisons comfortably in South Africa, if their behaviour and sexual proclivities are privately expressed. Since 1961 it has been held that sexual intercourse *per anum* is not an offence in itself, providing that it occurs between male and female parties. In terms of contemporary legal practice, it appears that the allowance has been tacitly extended to consenting male adults.[9]

Despite the foregoing, the situation of legal ambiguity has meant that homosexuals have come to believe that their behaviour warrants criminal sanctions, and hence they have *learned* to fear the law. An important clinical variable emerges from this, which is linked with the intermediate phases of homosexual identity development discussed earlier. This consists of the fact that homosexual internal oppression, associated with a long-term process of diminished self-esteem, has as its basis a residual anger directed at society and the legal system. This blocks the free expression of identity constructs, such as the spontaneous expression of love, the ability to share intimacy with family and friends, and the ability and security to be a politically viable person in the context of oppression. Thus a conscious defence exists, often blocking the sensations of same-gender bonding. Anderson (1982) as well as Hodges and Hutter (1974) examine this concept of internalized self-oppression. Hodges and Hutter remark as follows:

> We have been taught to hate ourselves — and how thoroughly we have learnt the lesson. Some gays deliberately keep away from teaching lest they be a corrupting influence. Others, except for brief, furtive sexual encounters, consciously avoid the company of gay people because they cannot bear to see a reflection of their own homosexuality (Hodges and Hutter, 1974:2).

This process, after due reinforcement from the sub-culture, for example by referring to the police in pejorative jargon (such as 'Priscilla'), becomes part of the well-established sub-cultural fear and anger towards legal authorities. By contrast, homosexuals in some European countries, parts of North America, and Australia, have long surmounted the fight for basic legal recognition. Their movements and alliances have new priority issues, such as homosexual marriages, tax rights, single and same-gender parent child custody cases, job discrimination, lowering of the age of consent to below 18 years, and political issues. For homosexuals in South Africa, this creates a form of envy; some feel impotent to effect change in comparison with their overseas counterparts.

With respect to South Africa, Helm draws attention to the differences between legal enactment and legal enforcement. She states:

> In the processes intermediate between the report or observance of a criminal act and the bestowal of legal status of criminal upon the actor, the police play a most important role. This role depends upon at least:
> (i) the individual officer's attitudes and interpretations of his role;
> (ii) the varying pressures which the community from time to time exerts on the police 'to do something' about the 'homosexual problem';
> (iii) the organization of the police department (Helm, 1973:10).

Upholding Helm's statement, information received from the Attorney General's office in Cape Town indicates that, since 1972, no case of sodomy between consenting adults in private has been prosecuted, and it is not policy to do so. In concert with the Attorney General's office, both the Cape Town Regional Magistrate's Court and the Wynberg District Magistrate's Court have no records of any prosecutions of consenting adults in private. The policy of the Attorney General's office is followed in this regard.[10]

Rueda (1982) and Babuscio (1976) provide illustrations of how, in the United States, police interpretations of community condemnation of homosexuality increase the likelihood of harassment and/or arrest. In South Africa, waves of arrests and swoops on gay people and institutions seem to be related to how the officer in charge of the vice squad interprets his role: in Durban (Natal) during the late 1970s and early 1980s, a particular vice squad commander 'cleaned' the city of homosexual clubs, bars, and cruising spots. Durban has only recently re-established its network of clubs and other meeting places in the wake of this spate of arrests and harassment.

An equally important phenomenon is evident in the 'new wave' of police attitudes in respect of upholding the law. Police have been known to respond with sympathy to people who have been attacked, molested, or 'bashed' by 'queer bashers'. A client seen professionally by one of the writers provides corroboration of this. He was attacked by a 'non-gay' person and badly beaten up, but disguised the facts of the case to the police, for fear of reprisals against him for being gay. After a short time, the investigating officer confronted the client by opening up the possible homosexual link between the attack and three murders which had occurred, the link being that the victims were all known gay men. Police co-operation, including attitudes that were not anti-gay, facilitated the speedy arrest of the accused, and triggered off a ripple of respect for police intervention.

The so-called co-operation between police and gays in some cases has been extended to actions in the lower and higher courts. Court proceedings in respect of gays who have been victims of assault clearly indicate that these cases have been viewed strictly on their merits, which implies that there has been complete recognition of the civil rights of the gay people who have been assaulted or killed. It would thus appear that the central consideration of the police (and the courts) is not so much that the victim is a homosexual, but that as a person, his basic rights of life, property, and security should be defended (*Argus*, 11 August, 1984; *Cape Times*, 22 April, 1986).

The formal gay movement in South Africa

It is with the foregoing discussion in mind that the historical development and present nature of the South African formal gay movement is described and analysed below.

The first semblance of a formal gay movement emerged in South Africa in January, 1966. The police raided a large party hosted in a private home in Johannesburg. Prominent white business people, advocates, doctors, artists, and people who had political connections were present. Although the raid took place under the Liquor Act on the premise that alcohol was being sold without a licence, nine arrests were made and charges were levelled against many people 'because of one or other [alleged] homosexual deed' (Joubert, 1974:2).

The raid and its consequences received widespread newspaper publicity. A group of homosexual people, including some of the party-goers, were strongly motivated to establish an action group to secure the release of arrested people, ensure their legal defence, and to protest against the police action at the party.

Possibly stimulated by the party and its aftermath, a draft Parliamentary Bill was proposed to introduce strict new amendments to the Immorality Act to 'control' homosexuality. This was controversial, both within and without Parliament, so that before proceeding with the Bill, a Parliamentary Select Committee was appointed to investigate homosexuality. The action group responded by extending its planning and fund-raising activities, in order to employ the services of legal officers and expert witnesses to give evidence at the inquiry. Retrospectively assessing the success of the group's endeavours, Joubert (1974) maintains that the action group's response was a determining feature of the collective evidence presented to the Parliamentary Select Committee.

The result of the Select Committee's inquiry was a recommendation to leave the existing law relating to homosexual behaviour unchanged, bar one issue: the age threshold for punishable offences with minors was raised from 16 to 19 years.

Hence, a group of people showed that by mustering their collective forces, both academically and financially, and entering the political process to present evidence to the Select Committee, it was possible to influence significantly the development of public policy. The action group and its supporters provided a forum for gay debate, and ultimately paved the way for aspects of the homosexual movement to develop in South Africa.

From a personal interview held by one of the writers with an organizing member of the original action group, it is clear that particular circumstances affected the group's impact and life span. The group had two very specific and pressing objectives: to bail charged partygoers out of prison, and to support their defence; and to engage the services of advocates and others to give evidence to the Select Committee. Once these objectives had been achieved, the impetus for the group gradually petered out. Another factor that contributed to the demise of the group was the lack of a formal structure. The group operated informally and semi-clandestinely, out of fear of public disclosure, reprisals such as the loss of jobs, and blackmail. This is probably also the reason why no regular ongoing records of activities were kept. The respondent to the personal interview consented 20 years after the Select Committee's Inquiry to have his name in print. 'I have no fear now of publicly sharing my homosexuality with others,' he said. 'Up until a few years ago my status in the commercial world and international reputation with a service organization prohibited my gayness. I am fully out of the closet now.'[11]

Following the disbanding of the action group after the outcome of the Inquiry in 1968, attempts in the early 1970s by gay people to maintain the impetus for law reform and action proved to be unsuccessful.

Hampered by lack of leadership and funds, and general mistrust in the gay community, the gay movement did not succeed in achieving much more than short-lived publications and spurts of meetings to discuss legal issues.

The publications of the time were either bland and fairly widespread, or relatively activist but with a small readership. An example of the first is the magazine *Equus*. This glossy Johannesburg magazine attempted to offer a forum for gays to share their feelings, to promote gay venues in Johannesburg, and to offer articles that included gay themes. The magazine also included pin-ups of local gay men. While *Equus* was probably the first attempt to offer the gay 'community' in Johannesburg a form of public identity, the contents were inoffensive and un-radical, so as not to offend the authorities. An example of a more activist publication was *Comment*, a monthly pamphlet circulated by 'Gays Anonymous', a group of concerned Christian gays in Cape Town who believed that Christian principles and gay liberation were compatible. Unfortunately a sense of anonymity pervaded the publication, leading to its demise after nine years. Nevertheless, the watchdog role played by people involved in *Comment* was valuable. This included regularly monitoring newspaper reports, arranging counselling, encouraging social interaction, and dealing with local homophobic issues.

Further *ad hoc* groups were established to publish on themes related to homosexuality during the 1980s. Examples of some of the publications include *Alternatives*, *Gay Between*, *Coming Out*, *Lambda*, *Young Ideas*, *Gay Christian Movement*, and *The 60-10 Newsletter*.

During the 1970s and early 1980s, gays made attempts to establish local gay organizations, and it was three of these organizations (Lambda, Unite, and Amo) which combined their 152 members together on 1 April, 1982 to form the nucleus of a new national organization, The Gay Association of South Africa (GASA). Branches were subsequently established in all main centres, with existing local organizations sometimes becoming chapters of GASA.

While GASA was established as an organization to unite all gays without discrimination to work together for a better understanding of 'the gay choice', it was manifestly un-radical. Its membership advertisements included sentiments such as the following: 'Remember, GASA is not a militant organization planning protest marches through city streets, and your membership will not imperil your privacy — all mail is posted in unmarked, sealed envelopes', and 'Join GASA and help to provide a non-militant, non-political answer to gay needs' (*Link/Skakel*, February, 1983).

The GASA newspaper, *Link/Skakel*, became the most widely read local gay newspaper in South Africa. It was subsequently replaced by *Exit*, which is probably the most successful South African gay publication to date. It deals with political matters of interest, local news items, and cultural and academic articles written by prominent gay people. Emphasis is upon highlighting the 'gay scene' in South Africa, publicizing gay resources, and reporting homophobic attitudes in the country. The content of the paper follows models in other countries, including *The Advocate* (USA), *IGLA Bulletin* (International Gay and Lesbian Archives), *Body Politic* (Canada), *Gay News* (United Kingdom), *Homokrant* (Holland), and *Siegessaule* (Germany). Some criticism levelled at *Exit* has included its sexist and homoerotic flavour, a seemingly biased sense of reporting when political issues are at stake, and a failure to address homosexuality from a leftist perspective.

As a result, splinter papers have mushroomed, and attempts to balance the 'conservative image' of *Exit* have emerged. Examples of some opposing pamphlets that are in circulation include *Lago* (published by the organization 'Towards a Progressive Gay and Lesbian Alliance'), the OLGA newsletter/pamphlet, and *The Congress of Pink Democrats* (published by an organization of the same name). The primary mandate of these publications is to reflect principles of non-racism, non-sexism, and non-collaboration.

Groups such as those represented by the above publications were originally members of GASA, which have broken away to form, in effect, their own alliance. A strong connection exists between Lago and the Pink Democrats with the newly formed progressive gay groups, such as the African Gay Association (Cape Town), OLGA (Cape Town), and the Rand Gay Association, which is now known as GLOW (Johannesburg-Soweto).

The levels of fragmentation and polarization of ideas in the gay collective are illustrated by a national pamphlet in which GASA (Johannesburg chapter) challenged the Rand Gay Organization to explain the stance on homosexuality of the African National Congress (ANC). GASA took the stand that before commitment can be made towards the wider political liberation struggle, clear protocols must exist in respect of the ANC's attitude towards homosexuality, which is ambiguous. Despite alluding to the need for social justice, ANC spokespeople have suggested that homosexuality is 'not normal' and that minority rights are irrelevant in the struggle for majority rule (*Capital Gay*, 18 September, 1987:11).

The polarization of political ideologies reflected above can be partly understood by reference to a local study conducted by Normann in 1983. Normann's study attempted to define the 'gay ideology' of groups of people who belonged to established organizations in Cape Town, including the 60-10 organization and The Gay Information Working Group on the campus of the University of Cape Town. His findings revealed two distinct levels of gay ideological thought.

In the first place there were those who perceived gay liberation as a process of interaction to facilitate the individual person coming to terms with homosexuality (in terms of definite role models). This implied rejection of 'closet' behaviour. People in this group also placed emphasis on legal reform, and on the education of society in sexual liberation, since they believed that the reform of laws on homosexuality would be ineffective without addressing social prejudices as well. The second level of opinion held that the countering of oppression in *all* forms should be a priority of Gay Liberation. Oppression must be viewed as a larger process, broader than only gay liberation. Moreover, people in this group believed that action should be confrontational, emphasizing societal and structural change (Normann, 1983:29–32).

Perhaps the GASA attitude mentioned earlier represents the first position identified by Normann, while that of GLOW (Gay and Lesbian Organization of the Witwatersrand) and OLGA support the second. Although much criticism can be levelled at Normann's study, particularly his biased sample of middle-class respondents, his study nevertheless offers a flavour of local opinion, and lends support to the writers' contention (discussed earlier in Chapter 4) that gay attempts at 'homosexual liberation' in South Africa are split.

Indications of the split are apparent in the following:

▶ The recent demise of GASA (National) as a result of political intrigue, poor handling of funds, and uninspired leadership.

▶ Political splits based on racial and political differences, which have resulted in, on the one hand, leftist offshoot organizations, and on the other a moderate reform majority, subscribing to the original aims of GASA. Within the leftist offshoot organizations, further splitting occurs because of separatist ideologies, such as political feminism in some lesbian groups.

▶ An inability to determine 'gay priorities', which have been variously identified as AIDS, legal reform, sport, service centres providing counselling, and political campaigns.

The eventual collapse of GASA (National) has paved the way for a preliminary investigation of the very issues raised in the first point above. Under the caretakership of an elected national chairperson and secretariat, GASA has reorganized its national structure, with a mandate to form a national Gay Alliance of South Africa. A major consideration proposed by the architects of the alliance is to decentralize the power base, and to encourage each region to develop its own independent gay social and political structures. Each centre will hopefully enjoy a sense of self-expression and therefore determine needs and priorities from a local base as opposed to a national one. The proponents of this structure anticipate that federalism will activate clusters of people who feel emotionally, socially, and politically safe to execute their own particular mandate within their localized perspective.

Insight into the lack of success in creating a unified gay liberation movement in South Africa can be gained from the work of Lee (1977), who has analysed the sociology of homosexual liberation in the United States. Lee uses a symbolic interactionist model, similar to that used by the writers in discussing homosexual identity development in Chapter 1, to isolate the dynamics that seem to influence the nature and success of gay liberation movements.

Firstly, Lee emphasizes that, unless homosexuals have passed through the stages of personal identification, including the 'ego-destructive' behaviour as noted by Humphrey (1972), attempts at a corporate identity (i.e. 'liberation') will prove futile. He states:

> The whole process can be considered as a 'status passage' or 'moral career'. The emphasis will be on process, in the symbolic interactionist sense, involving the concept of self as the reflexive agent, socializing itself into appropriate roles as well as fitting into the slots provided by society (Lee, 1977:52).

Secondly, Lee offers three variables which mediate the process of acquiring a 'corporate identity'. They are coming out, signification, and going public.

Lee's variables suggest that 'gay liberation' has, symbolically, its own identity growth protocols, and that they must be seen in the context of individual stages of coming out and identity growth. 'Coming out' identifies a *beginning commitment* to sexual identity, while 'signification' reflects a *significant commitment* to homosexuality. 'Going public' indicates an overall *ownership of identity* that transcends private commitment. 'Going public' occurs both in the sphere of sharing aspects of one's sexuality with others who are non-gay, and promoting homosexual ideology from a social action point of view.

South African gays, tarnished like all South Africans by exposure to apartheid divisions and oppression, and also made insecure by legal uncertainties, are for the most part far from Lee's stage of 'public ownership' of their identity. It follows that there will not be widely supported national efforts by gays to form movements which publicly and actively promote political reform and gay liberation. In different words, they will not form popular movements with an *external action* focus. South African circumstances are such that formal gay organizations will be more likely to have an *internal action* thrust, such as promoting the quality of gay life and meeting the everyday needs of gay people.

A brief overview of existing formal gay organizations in South Africa supports this contention. The organization of gays on a national basis is, as reported upon above, in a state of flux. Because of a multitude of factors — political, racial, organizational, and others — stable gay unity at the national level has thus far eluded South African gays. In consequence, there is no well-supported and established national body to promote gay interests, either internally within the gay collective, or externally in the wider society.

On the other hand, relatively well-supported organizations with predominantly service functions exist in all the main regions of South Africa. Services include counselling activities, sporting and cultural events, social gatherings, religious support groups, and student groups at certain universities. Service delivery — culturally, socially, or supportively — appears to have the endorsement of the gay collective, perhaps because these services contribute to the well-being of individuals.

Features contributing to crisis

The latter parts of this chapter have outlined the development and nature of the formal gay movement and formal gay organizations in South Africa, against the background of the particular South African situation. In the discussion, unique circumstances, problems, and paradoxes emerged, many of which contribute to a series of crises which influence the identity development of homosexual people. Some of the central ones are summarized below.

▶ *Generalized homophobia,* from both within and without the gay collective, serves as a deterrent to consolidated action by gays. Kenneth and Behrendt (1987) suggest that homophobia in both heterosexuals and homosexuals creates one of the major stress factors for gays, and that it has severe implications for the sources and resolution of conflict, particularly around the area of legitimacy.

▶ *Proscriptive legislation* against homosexual sex acts suppresses attempts to legitimize homosexual aspirations at the personal, social, occupational, and political levels.

▶ Generalized homophobia and proscriptive legislation have in combination influenced the coming out patterns of gay people. *'Closet behaviour'* is still the rule, rather than the exception. As such, homosexuality is represented by a visible minority but practised by an invisible majority of homosexuals.

▶ *Moral opposition,* including certain attitudes of organized religion, has promoted feelings of guilt and unworthiness in homosexuals. Linked to the power of the family system, the unavoidable consequences of rejection and abandonment plague many gays. In the words of Cramer and Roach, 'most relationships between gay sons and their parents are strained immediately following disclosure, and a period of turmoil usually ensues for most families ... which will lead to a family crisis' (Cramer and Roach, 1988:87–9). Thus there are many homosexuals who will forfeit the expression of freedom for ostensible family and/or societal acceptance.

▶ South Africa's *apartheid* history of racial oppression and black political resistance has created a complex socio-political situation that affects all dimensions of society, and sexual minority group issues are no exception. Gay liberation within doctrinaire apartheid will suffer the same consequences that voluntary groups, political groups, and others experience. The momentum and credibility of any form of gay liberation will be retarded in a politically oppressive society.

▶ The *lack of credible and vociferous public leadership* (including powerful lobby groups and international support) adds to the crisis of gay liberation in South Africa. South African gay leadership is less obvious in comparison with other Western countries, for the following reasons:

▷ leaders are not public with their identity;
▷ they do not all subscribe to the same ideology;
▷ they are often identified only with a particular region or a particular cause, and therefore are not representative of all gay constituents; and
▷ many are identified as 'leaders' not on gay ideological merit, but on the basis of their activities, such as being owners of clubs, organizers of competitions, or working with people living with AIDS.

However, some people have made considerable strides to place homosexual concerns in the wider socio-political arena, and have gained credible images as leaders within segments of the gay collective. Among these are prominent academics, lawyers, ministers of religion, business people, professional social workers, doctors, and trade unionists.

▶ *AIDS* has created a paradoxical situation in South Africa. While exposing homosexual proclivities in an unparalleled way to the public through the media, it has retarded the liberation force in two ways. Firstly, it has resurrected homophobia within the gay sub-culture and the wider parent culture, with emphasis once again on promiscuity and lifestyles that are seen to be contrary to societal mores and beliefs. Secondly, it has reintroduced a backlash of political innuendo (a) among heterosexual whites, who regard AIDS as a form of divine retribution, and who are consequently indifferent to the suffering of AIDS people, and (b) among blacks, who maintain that homosexuality is a 'white man's [sic] disease', with the implication that AIDS belongs only to a sub-group, racially and sexually.

▶ The sub-culture has as its emphasis a *hedonistic value system*, and seems incapable of collectively contributing to homosexual liberation or societal change. This captures the essential spirit of the crisis in the gay movement, in that the sub-culture is the most powerful contributor towards the process of gay identity. Furthermore, the sub-culture has not yet reconciled its needs with the needs of the broader South African society.

Notes

1. Full details are provided in J Lauritsen and D Thorstadt, *The Early Homosexual Movement (1864–1935)*, New York: Times Change Press, 1974.
2. Detailed analysis of the divergent Australian experience is offered in Chapter 8.
3. See for example D Altman, *Coming Out in the Seventies*, Boston: Alyson Publications, 1981; J Chesebro (Ed.) *Gayspeak Male and Lesbian Communication*, New York: The Pilgrims Press, 1981; T Marotta, *The Politics of Homosexuality*, Boston: Houghton Mifflin Company, 1981; L Richmond and G Noguera, *The New Gay Liberation Book*, Pala Alto, Ca.: Ramparts Press, 1979.
4. For an in-depth review and analysis which highlights the divisive structure of South Africa, see E Boonzaaier and J Sharp (Eds.), *South African Keywords. The Uses and Abuses of Political Concepts*, Cape Town and Johannesburg: David Philip, 1988.
5. In its 'divide and rule' strategy over Africans, the Nationalist government legislated for each tribal grouping to occupy its own reserve or Bantustan, for example Tswanas in Bophuthatswana, Zulus in KwaZulu.
6. Although homosexuality is forbidden in the Muslim fundamentalist context, it is practised by a sizeable proportion of 'coloured' gays. The following two points pertaining to religious opinion have been noted by one of the writers in the course of his professional interaction with Muslim gays:

▶ The *idea* of being homosexual is confusing for some Muslims, particularly within a patriarchal religion and culture. In essence, homosexual *behaviour* is seen to be 'effeminate, emulating the female gender'. This poses a problem for worship, in that women follow men in prayer, and are segregated as well. Thus a man who is 'partially female' cannot lead women in prayer, and his place in worship is therefore arbitrarily defined as 'non-legitimate'.

▶ Two Muslim men who had been in a relationship for 12 years needed therapy to mediate a domestic conflict. Part of this conflict arose from the concern of their families and friends (all of whom were Muslim) about a potential rift in the relationship, highlighting the sacredness of a union between two people, irrespective of their gender. Furthermore, the older of the two partners was a devotee of the Faith and had unconditional support from his spiritual community. The crisis was in effect exacerbated by the pressure of families and friends, who believed that the relationship was cemented by a sacred and legitimate union, and that the commitment had to continue.

7. This Act has been repealed and the sections relevant to homosexuality are now contained in the Sexual Offences Act of 1988. The legislation affecting homosexuals remains unaltered.
8. Under common law, sex between men is criminal, and is punishable in both civil and military courts (P M A Hunt, *South African Criminal Law and Procedure*, Volume 2, second edition, Cape Town: Juta and Co, 1982:271–8).
9. Anal sex is in many instances a preferred form of sexual intimacy between gay men. This has been supported by post-AIDS literature, especially those studies concentrating on epidemiology.
10. Personal communication from the Attorney General's Office.
11. Personal interview with Mr J Garmeson.

7 Homosexual identity formation, culture, and crisis: an empirical study

Research into homosexual issues

There has been considerable research into homosexual behaviour during this century, but this enquiry has not moved steadily along a single thematic path. Indeed, a perusal of research studies over the last 50 years reveals a significant change in the focus and nature of research from the period preceding the 1970s, to the period afterwards.

In 1974 Weideman published an annotated bibliography of 1 265 publications on homosexuality which had appeared in the English language between 1940 and 1968. Scrutiny of these contributions reveals that many of them sought to explain homosexual behaviour in terms of congenital and psychological pathology. By contrast, post-1970s research veers strongly away from the pathological approach, and seeks to examine homosexuality through a social microscope.

In 1984, Shively, Jones, and De Cecco reported that since 1969 they had been able to locate over 1 100 unduplicated articles on homosexuality from 47 scientific journals. From this universe, they selected 228 items for scientific scrutiny in terms of the general methodology employed and the approach used to define sexual orientation.

The most common methodologies were field studies and psychometric investigations (37% and 29%, respectively), with the balance made up of a range of other research approaches. Experimental studies using a control group, which are the only type of research design that can scientifically demonstrate a cause-effect relationship, comprised less than two per cent of the total.

In terms of definition of sexual orientation, nearly three-quarters of the studies surveyed defined this operationally i.e. the presence of homosexual *behaviour(s)*. In half of the studies, physical sexual activity was the only definitional element, with other aspects such as affectional fantasies, close relationship arousal, and erotic preference being used far less frequently.

Shively and his associates therefore conclude that in most studies, sexual orientation has been treated as if it was a perceptible, unitary phenomenon (Shively et al., 1984), rather than acknowledging that sexual identity and sexual orientation are neither unitary nor mutually exclusive.

Thus, many methodological and conceptual limitations mar even relatively recent studies on homosexuality. Nevertheless, much modern research is characterized by a broadening of conceptualization along the lines suggested above by Shively and his colleagues, and a widening of research focus. There has been a recognition that homosexual behaviour is not necessarily dependent on sexual conduct. It can include non-sexual areas of awareness, such as consciousness on the part of homosexuals that they constitute a sexual minority, that sexual identity has a symbolic content (manifested in values that are constantly under attack), and that they belong to, or belong on the periphery of, a gay sub-culture. Moreover, gay identity has appeared as a political discourse on sexual behaviour, and the gay collective has been examined. There has been a recognition that idiosyncratic sets of individual behaviours and opinions cannot be studied on their own, and that they have to be seen in relation to the socio-cultural manifestations of sexual identity.

Research design and methodology

Aims and objectives

In earlier chapters of this book, the view has been presented that achieving a homosexual identity is an ongoing process. The identity does not derive from sexual conduct only but includes identification (or lack thereof) with the gay sub-culture and its various facets. It has also been suggested that identity growth is not synonymous with the gay argot expression of 'coming out'. Coming out, or owning to aspects of identity, is merely one feature of identity acquisition. Moreover, it has been postulated that the phases or stages of homosexual identity development are influenced by successive crises, and that these crises may be experienced as a negative or as a growth-promoting experience.

The *aims* of this study are to establish whether or not empirical support exists for these postulations. Hence the study has two general aims:

▶ To gauge the impressions which male homosexuals have of the development of their sexual identity, by eliciting retrospective responses to their perceptions and experiences of same-gender attraction.

▶ To establish whether or not respondents were able to identify their homosexual growth, or parts thereof, as being influenced by a crisis.

The foregoing aims can be operationalized more specifically by examining the extent to which empirical evidence supports (or does not support) six suppositions, which represent the main themes of earlier discussion in this book. These are:

▶ Homosexual identity development is a complex and on-going process. Homosexual unfolding is experienced in varying degrees and at different stages during the life of an individual. It is never a single, definitive event that suddenly occurs.

▶ Identity development is influenced by the presence (or absence) of crises. These crises have the potential to stimulate pathogenic behaviour, or growth and development.

▶ Identity development takes place through a series of cumulative incidents, which have both cerebral and behavioural components. If any of the incidents are not successfully negotiated, the residual components of that incident can manifest at later stages and retard (or accelerate) the identity formation process.

▶ Homosexual identity development begins at an early age, becomes definable at puberty and, in taking its course, is affected by sub-cultural assimilation or rejection.

▶ 'Coming out', or owning aspects of homosexual identity, is not synonymous with identity development, but only a part of it.

▶ The homosexual sub-culture, which is influenced by a tradition of sub-cultural forms of expression and behaviour, produces a double-bind effect. It metaphorically procreates homosexuality, while at the same time binding a homosexual individual to a marginal sub-culture. The gay sub-culture has an allegiance to its own survival, and thereby it indirectly fosters gay self-oppression.

Design of the study

The present study is a field survey — a survey, by means of an anonymous mailed questionnaire, of the responses of a group of homosexual men. Essentially, the study is concerned with the internal frame of reference of respondents, for, as Cass notes, 'After all, it is the individual's own perceptions of the world, rather than the world itself, which are critical to the identity issue' (Cass, 1984:112). In the descriptive analysis of responses, use is also made of material from the writers' clinical practice, since these experiences and illustrations often help to clarify and concretize a point made. While the use of clinical experiences and case excerpts involves subjective selection, these illustrations often provide a depth of insight hard to obtain by other means. As Lee observes:

> Social science has gained and will continue to do so from many types of contribution by social scientists of various sorts, but it stands to grow most substantially from clinical study. This type of study is the concerned, objective, intimate, continuing and thoughtful observation and critical evaluation, and absorption into evolving theory of spontaneous social responses to corrective or manipulative efforts. A given effort is justified in many ways. Justifications, responses, and all other observable aspects of such study becomes [sic] basic data for its pursuit. Only clinical situations thought by participants to involve social issues important to them are likely to yield significant contributions to social science (Lee, 1966: 330–1).

Research process

A *questionnaire* was constructed, comprising 20 closed-ended questions (with additional opportunity for commentary). A *pilot study* was undertaken, where the questionnaire was administered to 10 informally recruited homosexual men who were not members of any formal gay organization. The results indicated that operational definitions of crisis and homosexuality were necessary in order to reduce the ambiguity of key words and phrases. It was also possible to tighten the questionnaire used in the pilot study by reducing the number of questions to 13 and focusing them more directly on crisis.

Sampling was a major problem, largely because it is not possible to obtain a representative sample of homosexuals. First, homosexuality, as Gochros (1984) and Isaacs and Miller (1985) have shown, is not a fixed and static state, because of the variable identity status of people over time. Second, the full universe of homosexuality and homosexual behaviour cannot be established. Most homosexual behaviour is not visible, and what is visible (by action, statement, or symbol) represents only a fragment of the homosexual collective. Permission was thus obtained from GASA 60-10, a gay service organization in Cape Town, to administer the questionnaire to its 200 members. Hence, unlike the purposive sampling used in most previous research in South Africa, the present study made use of a non-probability sampling procedure. Such a procedure is justified when precise identification of the total population is not possible (Grinnel, 1981:86), and where the 'self-selection' of members joining a particular organization forms part of the sampling process.

Nevertheless, the sampling procedure used introduces a distinct bias into the study. Membership of GASA 60-10 may appeal to people who nominally identify with the gay collective, but the service organization is not associated with 'mainstream' gay culture as exemplified by bars, clubs, and so forth. In addition, the confidentiality stressed by the organization may appeal to people who feel unsafe or unsure about their identity, and who prefer to interact within the parameters of the organization's activities, rather than face the onslaught of the gay sub-culture. Finally, membership of the organization is open to female homosexuals and any person who is 'gay sympathetic'. The limitations of the sample are further discussed in a subsequent section on the limitations of the study.

Once a final decision about the sample had been decided, a letter of motivation and assurance about individual confidentiality was compiled in collaboration with the chairperson of GASA 60-10. This letter was sent to each member, together with the questionnaire and a set of definitions of key terms. Three months later, a letter of reminder was sent to each member, reporting the good response thus far and appealing for further replies. After a month, a cut-off point was decided upon. The responses then numbered 90 (i.e. a 45% response rate from members had been achieved).

The research tool

Two matters about the research tool require clarification: *why* a questionnaire was chosen, and what its *content* was. It was decided to use a questionnaire (as opposed to a face-to-face interview) because it enables full anonymity, because access by mail is less threatening than a personal approach, because it enables standardized wording of research questions, and because the element of interviewer bias is removed.

The questionnaire (Appendix B) contained 13 questions, preceded by a request for brief identifying data such as age, area of residence, and occupation. Only two questions (1 and 2) were open-ended, while all the others provided a closed selection of possible answers (although in addition they provided room for any commentary the respondent wished to make).

Questions 1 and 2 were deliberately selected to elicit recall responses pertaining to the unfolding of homosexual identity. Question 1, which asks for the approximate age of onset of same-gender attraction, was designed to yield data vital to understanding the genesis of homosexual feelings. Question 2 is a relatively standard question asked of homosexual respondents as a means of ascertaining their idiosyncratic responses and/or experience relating to the incidents/events which gave rise to more definite feelings of homosexual orientation. Coming out, in homosexual experience, is traditionally regarded as a milestone in composite homosexual behaviours.

Questions 3 and 4 sought to ascertain a response to 'crisis'. While the specific nature and type of crisis was not asked for, the provision for additional commentary gave opportunity for this.

Question 5 blends in with the first two questions, since it concerns the process of coming out. Awareness of this phase of identity can be blurred with general homosexual interpretations, as opposed to specific personal feelings. Hence the aim of the question was to elicit as many variables as possible in order to determine the efficacy of the response. Also, since this question is related to the first two questions, the answers could be of assistance in picking up contradictions.

Questions 6 and 7 were 'deflecting' questions, designed to promote response about the generic aspects of crisis, as well as defusing any feelings of guilt or discomfort about 'coming out'. Question 6 was intended to yield information about one of the specific themes of the research study, as it was suspected that 'other crises' were as important as 'coming out' in the transitional experiences of homosexuality. Question 7 served as a projective device, for acknowledging crisis in others is safer than admitting to self-crisis. It also helped to extend the theme of crisis beyond the personal experience of the particular respondent.

Question 8 was a seminal question. It was framed to reflect the sense of 'difference' or apartness that homosexuals allegedly feel, as well as to deal with the sub-cultural value of separateness. Since it was hypothesized that a sense of difference or separateness would bind a homosexual to the sub-culture, the existence of such feelings could help in achieving identity. Thus the question aimed to tap indirectly sub-cultural responses without referring to the sub-culture.

Question 9 has literal and figurative dimensions. Literally, the question was designed to gauge attitudes towards the homosexual service organization (GASA 60-10). Figuratively, two other considerations were probed: (a) homophile responses to an exclusive homosexual organization, thus furthering the issue of separateness; and (b) placing GASA 60-10 within the ambit of the sub-culture, or validating homosexuality and homosexual concerns through a formal organization.

Question 10 followed through on the theme of the GASA 60-10 organization, dealing more specifically with identity issues and sub-cultural aspects. One of the alternative responses was directed towards a sense of disappointment with the organization. By choosing such a response, a person could indirectly reflect the fear commonly expressed by gays that a formal organization does not cater for their sexual needs. It has been alleged that once a person feels this, he will tend to relinquish membership and participation in the organization. Finally, the word 'snobbish' was used in one alternative response as a euphemism for 'old queens', 'up-tight', and 'boring'.

Question 11 asked about feelings towards others in distress. Superficially, it reflects altruism, but it has other dimensions, too. It was also designed to reveal aspects of narcissism (the need to indulge in aspects of the self that are gratifying), while simultaneously containing a projective aspect. Helping others in distress may either reduce the helper's own distress, or distress personally experienced may facilitate empathic help for others. This question additionally has the potential to tap the localized response that 'gays are distressed people'.

Question 12 links directly to the previous question by qualifying the earlier response and associating it concretely with GASA 60-10.

Question 13, the final question, sought to establish respondents' feelings about being part of a research programme. Positive or negative feelings could also be symbolic statements of respondents' sense of homosexual resolution.

Limitations of the study

There are four major limitations of this study of which readers should be aware, and which they should bear in mind when they evaluate the study findings:

- ▶ Since the universe of homosexual people could not be defined, a purposive sample of respondents could not be drawn. The non-probability sample used (the membership of GASA 60-10) captured only a small, mainly white, middle-class segment of the gay population, which cannot be regarded as typical of all gays. Hence the size and nature of the sample prohibits use of the techniques of inferential statistics and the generalization of findings.

- ▶ While use of a questionnaire as the research tool has many advantages (discussed earlier), it also has shortcomings. For example, inflexibility in question format, lack of control of the environment, inability to probe answers, and the possibility of unanswered items and biased responses. It also has high potential for a low response rate. Despite these aspects, the advantages of the questionnaire outweighed its disadvantages in this study, and the appropriateness of its use is perhaps reflected in the exceptionally high response rate of 45%.

- ▶ The questionnaire did not directly address AIDS, which has subsequently become a major crisis for many homosexuals.

- ▶ Criticism may be levelled at a study based largely upon 'recall'. In this respect Hoult writes:

    In addition to the problems of faulty recall, there is the all-too-human tendency to reconstruct our autobiographies in an effort to bring them into greater congruence with our present identities, roles, situations and available vocabularies (Hoult, 1984:143).

    While Hoult's reservations undoubtedly have some validity, it must be noted that he made his comments in respect of studies into homosexuality dating back to the sixties, where recall was often determined by respondents' fear of revealing delicate and controversial aspects of their sexuality to uninformed and homophobic researchers. In this study, the homophobic element was absent. Moreover, research questions requiring recall were designed to elicit the 'sensation' of past experiences and feelings, rather than attempting to quantify them. Recall in these circumstances should not be underestimated, for it is the basis for determining the historical precedents which influence present behaviour.

The findings, and discussion of them
Age, residence, and occupation of respondents
Age

Table 7.1 Age of respondents

Age group (in years)	Respondents	Percentage
15–19	1	1
20–24	8	9
25–29	20	22
30–34	15	17
35–39	15	17
40–44	12	13
45–49	5	6
50–54	2	2
55–59	2	2
60–64	0	0
65 and over	2	2
Unknown	8	9
Totals	90	100

The age of respondents ranged from below 19 to above 65 years. There were respondents in almost every age group between, but with a clustering (84%) between 20 and 49 years. The median age was 30 to 34 years, and the mean age was 35. This age distribution corresponds with that in the benchmark Weinberg and Williams' study of 1975, which was reaffirmed by Berger's study of 1982. This finding points to a characteristic of the gay collective noted as early as 1948 by Kinsey and his associates, namely that homosexuals appear in every age cohort and in all social groups.

An interesting feature is that almost a tenth of respondents 'missed' the age question in the questionnaire, or avoided giving their age. McWhirter and Mattison (1984) note a similar phenomenon in their study, and speculate that older men in particular may have difficulties in accepting their age in contrast to younger people. This postulation is supported in the present study, particularly when the relatively low response rate in the age categories above 50 years is noted. Two possible explanations may be advanced.

Firstly, more mature people might be settled in a relationship, and/or leading a satisfactory, secure social life. This might make them less likely to join a gay association. Secondly, feelings and experiences of alienation might influence more mature people not to join an organized gay group. Alienation could arise from a fear of competition related to sexual prowess, a perceived decline of physical attraction, or reduced self-esteem; inhospitality directed towards them by younger people; rejection or ostracism, for it is not uncommon for older people to be pejoratively referred to as 'old queens'; and sensory overload, whereby it becomes too traumatic to deal with a collection of younger people who might trigger off the older person's sexual fantasies.

Forces such as these may place older people on the *periphery* of the sub-culture. For example, during the expansion of GASA 60-10, many attempts were made by some of the younger members to exclude more mature people from some of their social activities. A discotheque was started which overtly discriminated against older people, and a series of newsletters entitled *Under Thirties* was circulated, creating splinter groups in the organization. More recently, a pamphlet series entitled *Young Ideas* emerged as another vehicle of discriminating against older people.

The work of Berger (1982a, 1982b), Gould (1972), and Harry and DeVall (1978) underscores the detrimental effect on older people of contemporary age-conscious society, and of age-associated prejudice within the gay context. While age in and of itself is only a single variable, it is an important one (McWhirter and Mattison, 1984:159). The issue is, however, complicated by another consideration: if age has disadvantages for gay people, in which respects are these disadvantages different from, or additional to, the disadvantages that age brings to everybody within the relevant culture?

Residence

Table 7.2 Respondents' areas of residence

Area	Respondents	Percentage
South Atlantic seaboard	21	23
Central Cape Town	23	26
North Atlantic seaboard	2	2
Northern suburbs	8	9
Southern suburbs	28	31
Southern peninsula	6	7
Stellenbosch	2	2
Totals	90	100

The data in Table 7.2 indicate an urbanized set of respondents.[1] The majority (80%) live in an area within or adjacent to the Cape Town metropolis, confirming earlier research findings which point to the tendency for gays to live in or near urban complexes which provide socio-sexual outlets such as bars, discotheques, and community facilities. Places of residence include suburbs along the Atlantic seaboard from Mouille Point to Hout Bay, and suburbia referred to as the 'southern suburbs', which incorporates, among others, Rondebosch and Wynberg.

The concentration of urban and suburban gay people within the Greater Cape Town area does not constitute the ghetto-style settlements prominent in some North American cities (Levine, 1977). This is because a concentration of gay venues is characteristic of ghetto living and, relatively, there is a dearth of these in Cape Town. Nevertheless, clusters of gay people do live in some apartment blocks, which then acquire such nicknames as 'Queens Court' and 'Moffiehof'.

Two respondents resided in Stellenbosch and eight in Bellville. Both these places are some distance from Cape Town, and the residents are generally regarded as conservative and Afrikaans-speaking. An anomaly exists in the small response rate from these two areas, for both are known to have 'considerable' gay populations. The University of Stellenbosch has an informal gay organization, and gay coffee bars are located in the Bellville-Parow area. A shift in the location of social activities towards Bellville-Parow-Goodwood marks an attempt to decentralize the monopoly which the city holds over entertainment. Once a month a social ball is held at the Goodwood Show Grounds, and draws crowds in their hundreds. Part of this shift may also be a symbolic defiance of the multiracial integration that is growing in the Cape Town area, as the Goodwood activities are patronized primarily by Afrikaans-speaking people.

Note is also merited of the striking absence from Table 7.2 of people living in 'coloured' and African residential areas, such as Mitchell's Plain, Athlone, Bonteheuwel, Lansdowne, Grassy Park, Langa, Nyanga, and Gugeletu. Most of the respondents who completed the questionnaire were white. GASA 60-10 is treated with scepticism by many African and 'coloured' gays, and by those who subscribe to the politics of the 'gay left'. Hence splinter groups have mushroomed in the black areas, with little or no link with GASA 60-10.

A final characteristic of respondents in relation to their place of residence can be identified: their residential mobility. At least one third of the respondents indicated in their questionnaire that they intended moving to another area or another city. Relatively frequent address changes are common among the homosexual population. Geographical mobility can be an indirect consequence of identity conflict (often associated with closet behaviour), and/or the mismanagement of relationships (which ultimately has a bearing on identity). Mobility frequently occurs when relationships collapse, and an ex-partner is left homeless. In stable relationships, too, one partner may move when the other is transferred or obtains a better job elsewhere.

Occupation

Table 7.3 Occupational status of respondents

Occupation category	Respondents	Percentage
Professional	39	43
Commercial	15	17
Administrative	7	8
Skilled	9	10
Student	13	15
Unemployed	3	3
Retired	3	3
Unknown	1	1
Totals	90	100

In analysing the occupational status of the respondents, use is made of the simple three-category model employed by Berger (1982a:136), namely 'high status' (e.g. business executive, professional), 'medium status' (e.g. administrator, small business owner, clerical worker), and 'low status' (e.g. machine operator, unskilled labourer). In terms of Berger's paradigm, all of the respondents known to be gainfully employed were occupying medium or high status positions, with high status positions predominating. The latter included doctors, lawyers, university professors or lecturers, dentists, accountants, and architects, among others.

These findings have a significant bearing on the issue of 'going public', as described by Lee (1977). Self-disclosure is often presented as if it were a panacea for both the personal integrity and social progress of gay people. However, it is common knowledge that the possibilities of promotion or the chances of job satisfaction can vanish as the result of a homosexual label.

Employment discrimination exists against people who are unable or unwilling to conceal their homosexuality. In the United States, as far back as the Kennedy administration, the federal government began to wage a campaign to protect minority groups (including women) from discrimination in employment. Subsequently, Congress enacted comprehensive legislation to this effect and, according to Knutson (1979:173), 'many local governments and agencies have adopted ordinances and regulations which seek to protect the employment opportunities of gay persons'. In South Africa, in the absence of such legislation, discrimination at work usually results in non-disclosure of homosexual identity, which has serious consequences for the process of identity consolidation. Hence, gay identity may be hampered or advanced according to the choice of occupation and the social environment at work.

Choice of occupation may be consciously or unconsciously influenced by the need for compensation. As a result of being victims of discrimination, gays may be drawn to service careers such as medicine, social work, or teaching, which enable them to deal with feelings of alienation through being needed by others. This restores their sense of self worth. In the teaching profession in South Africa, however, any sense of self worth through effective performance is counter-balanced, especially in state schools, by fear of sexual identity being 'found out', and consequent dismissal. Another compensatory mechanism in occupational choice, affecting gays who lack self-esteem and who fear a lonely and isolated future, is to opt for high-profile positions which enable them to accrue wealth and possessions.

Certain occupations do not discriminate against homosexuality, thus facilitating a sense of *occupational freedom*. This minimizes the strain of dual identities ('straight by day, gay by night'). In South Africa such occupations include hairdressing, fashion design, professional theatre, male nursing, and lecturing at certain universities. A recent and interesting shift in this area has occurred in the state airways (SAA) where, until the AIDS panic of 1985, many flight attendants were openly homosexual. Now, flight attendants are required to be tested for HIV antibodies, and furthermore there is a seeming decrease in the employment of male cabin crew.[2]

Prejudice in many work situations inhibits disclosure of homosexual orientation, leading to *social isolation*, which in turn impairs expression of the self. This may account for the large number of instances of 'work paranoia' among responses to the question on 'other gay-related crises', discussed in full presently. The majority of respondents indicated that a fear of job dismissal or personal jeopardy existed in the work situation. This led them to be secretive and resentful. One respondent commented: 'In the space of two years I lost two consecutive jobs when my employers found out that I was gay. I felt like a fraud, and that my whole world had collapsed.'

Lee (1977) draws attention to the behavioural differences between 'coming out' and 'going public', and warns that while the terms are sometimes used interchangeably, they mean different things. Weinberg (1978) similarly distinguishes between 'doing' and 'being' gay, the latter suggesting a breaking down of social isolation, leading to the development of public ownership. The non-disclosure of sexual identity because of discrimination in the work place will maintain a person's sense of incompleteness.

In this regard Berzon (1979) proposes careful vocational planning, which includes ritual events that calibrate the gay existence. This would include disclosure (at some level) to work colleagues, as well as friends and family. The debilitating features of a dual existence and 'closet' behaviour should be emphasized, since they foster a form of emotional fraudulence that perpetuates a sense of prolonged or incipient crisis. In this study, the mean age at which respondents 'came out' was 22 years, yet analysis of their commentaries indicates a persistent fear of disclosing their identities in the work place. This signifies that part of their identity synthesis is still incomplete.

Of vital importance to the question of self-disclosure at the work place is the AIDS scare. Despite the fear of identity disclosure and a lack of support systems, many people have to *disclose prematurely their identity* to employers for reasons such as compulsory medical aid, which requires information about sexual risk behaviour; housing loans and life insurance, which require information related to sexual expression; and sickness, which, when combined with knowledge about HIV infection and AIDS, gives rise to suspicion on the part of employers. An example from one of the writers' case records reflects the last situation well:

> A person was diagnosed as having AIDS, not because of the illness with which he presented (an opportunistic pneumonia infection), but because of the friends who visited him. The nursing staff perceived them to be 'effeminate' and 'gay', and reported their suspicions to the physician, who then ordered a test without the patient's consent. Upon diagnosis, he was immediately discharged from his place of employment.

First recognition of same-gender attraction

Table 7.4 Age of respondents at first awareness of same-gender attraction

Age in years	Respondents	Percentage
0–4	2	2
5–9	19	21
10–14	33	37
15–19	26	29
20–24	7	8
25–29	3	3
30 and over	0	0
Totals	90	100

Two-thirds of the respondents first became aware of attraction to people of their own gender when they were aged between 10 and 19 years (mean age 13 years). Two respondents claimed to have experienced an attraction as early as infancy (0–4 years), but such early awareness is not uncommon (Colgan, 1987; Hetrick and Martin, 1987; Singer, 1981). A further 21% of respondents gave their age at first attraction as between five and nine years, with the age of seven recurring frequently. On the other hand, three respondents experienced first awareness as late as 25–9 years. Such late recognition is also not uncommon for some homosexual people, and is often the result of sexual fantasies and desires being deeply repressed, and surfacing on exposure to the gay sub-culture. Overall, the age distribution of respondents at the time of their first awareness of same-gender attraction closely approximated the findings of other studies, such as those of Coleman (1972), Daher (1977), Dank (1971), Hart (1984), Hetrick and Martin (1987), McWhirter and Mattison (1984), and Malyon (1982b).

The finding that most men experienced their first awareness of attraction to other males when aged between 10 and 19 years corresponds with the model of homosexual identity development outlined in Chapter 1, and confirms early and late adolescence as the most significant periods in the process of identity actualization. Tyson, another researcher, has also described adolescence as a critical time for the development of core identity gender, which is the precursor to gender role identity and sexual partner identity (Tyson, 1982:61).

In Chapter 1, acknowledgement of the first awareness of homosexual attraction was referred to as the 'homosexual sensation'. In their study into homosexual sensation, Roesler and Deisher define it as a significant series of pre- and post-puberty sexual experiences, regarded on recall as homosexual activities (such as mutual body exploration, masturbation, and fantasies), but without the connotation of 'being homosexual'. The homosexual experience (or 'sensation') usually precedes the individual's self-designation (Roesler and Deisher, 1972:1 018–19). Hence homosexual sensation should not be confused with homosexual identity, although they are associated. The process linking the two is 'coming out', discussed below.

'Coming out' and its association with crisis

The nature of coming out
Coming out was defined to respondents in the following way:

> By coming out is meant your recognition that you are homosexual, rather than heterosexual. This may happen via an event or situation, or an accumulation of feelings and/or experiences, whereby you admitted to yourself (and possibly others too) that you are gay.

This simple and practical definition does not contain within it some of the subtleties and complexities of the coming out phenomenon, and identification of these is necessary before turning to the findings of the research study.

First, a distinction must again be emphasized between the concepts of 'homosexual experience', 'coming out', and 'homosexual identity'. *Homosexual experience* concerns the cerebral and physical components of homosexual sex-related behaviour. *Coming out* usually relates to a set of experiences, and the consequent interpretation of them by the person as 'legitimate'. It involves recognition by the person that he is 'homosexual', and it is associated with the external rituals of the sub-culture which impact on the identity of the individual. *Homosexual identity* is the culmination of a formative process that has occurred over time.

Tripp (1975) proposes that the delicate line distinguishing coming out from other aspects of homosexual identity can be discerned by considering the systems of denial which operate when people practise homosexuality without having to admit to themselves or to others that they are homosexual (or gay), even though they may be exclusively so. Tripp further suggests that self-acceptance relates directly to the degree of denial of, or commitment to, homosexuality. Commitment increases a person's need for adaptive mechanisms, including ways of protecting himself from rejection (Tripp, 1975:139). In support of Tripp's contention, two case illustrations from clinical work are cited, which outline the coming out process in late adulthood (i.e. over 55 years).

> A sixty-seven-year-old retired professional presented with severe depression, fragmentation of the personality, suicidal thoughts, and a degree of psychosis with paranoid features. Upon assessment, and after history-taking, it was apparent that he was no longer able to control his homosexual fantasies, which had been repressed from an early age. Although the client was divorced, with adult children, he was unable to reconcile his fantasies (internal structure) with his external perceptions of how he believed others saw him (i.e. as a heterosexual). Some precipitating factors included retirement with excess time on his hands, and a deliberate need to behave like a voyeur on the beach and to obtain books with pictures of naked men. Therapeutic intervention revealed a person who for many years had suppressed his 'raw desires' and fantasies for fear of exposure or rejection.
>
> When his immediate family were informed of his feelings and consequently accepted his fantasy lifestyle (he had never formed a homosexual liaison), he had to be immediately hospitalized for fear of complete personality disintegration. Their acceptance of his alleged homosexuality had triggered off a wave of uncon-

trolled anger and resentment in respect of years of 'lost experiences', and he furthermore believed that he was unable to process and actualize his homosexual ideals. Ongoing therapy helped to contain his coming out fantasies and to place them in the perspective of his real world. Opportunity was given to him to mourn retrospectively his lost experiences and to deal with his anticipated sexual outlets in order of priority.

A teacher in his late fifties 'suddenly' told his wife and family that he was gay, left home, and began to network with homosexual people. Within the space of three months he had found a lover and began to frequent gay social events and private parties. His coming out process, although occurring in his later adult years, had thus far proved to be rewarding and meaningful. However, he resents his attachments to the heterosexual world with which he still maintains strong links, especially in his work situation. He believes that he will not openly divulge his 'new' identity until he has retired.

The above vignettes clearly illustrate two of Tripp's tenets. In the first instance, the crisis was directly related to the system of denial that had been in operation for many years. In the second instance, the person's degree of commitment to his homosexuality facilitated his identity resolution.

Cass (1984) and De Cecco and Shively (1984b) add a further dimension to Tripp's thesis by emphasizing the degrees of difference between homosexual identity and sexual identity. In their view, for many homosexuals coming out is seen to be an identity construct related solely to sexual acting out behaviour. A pattern of sexual behaviour is thus mistaken for homosexual evolution. Roesler and Deisher (1972:1 023) cite a psychiatric interpretation in this regard, and refer to sexually fixated behaviour as 'homosexual ideation'.

Berger (1982a) emphasizes the lack of uniformity in definitions of coming out, and suggests three factors that are related to this event: the first sexual encounter; openness with others about one's sexuality; and self-recognition and self-acceptance. The essence of coming out, according to Berger, lies not in these steps but rather in an acknowledgement of the overall process as 'the most significant life event in the experience of the person' (Berger, 1982a:22–3). Roesler and Deisher (1972) also stress this life event as being one of powerful significance. The root of the experience is that *coming out implies admitting that one is a homosexual.*

Age at 'coming out'

Table 7.5 Age of respondents at 'coming out'

Age in years	Respondents	Percentage
10–14	3	3
15–19	28	31
20–24	32	36
25–29	16	18
30–34	3	3
35–39	3	3
40–44	0	0
45–49	2	2
50 and over	0	0
Unknown	3	3
Totals	90	100

It is apparent from the above table that the majority of respondents (88%) acknowledged that their coming out process took place from mid-adolescence to young adulthood. The mean age for coming out was 22 years. Only three respondents noted that they came out when aged between 10 and 14 years, and eight in their thirties and forties.

This pattern corresponds broadly with the findings of a number of studies dealing with coming out (Berger, 1983; Coleman, 1982; Gadpaille, 1980; Humphreys, 1972; Lee, 1977; Roesler and Deisher, 1972; and Rueda, 1982). However, the mean age of 22 years for coming out is higher than that indicated by some other studies (Dank, 1971; Hooker, 1965; Jay and Young, 1979; Lee, 1977; McDonald, 1982; and Warren, 1974). This could be attributed to the fact that laws affecting homosexuals in South Africa are ambivalent and repressive. Identity is a social construct, and its variance is determined by social requisites and norms. Unlike the North American situation, where the homosexual liberation movement contributes to the coming out process (in the sense that coming out becomes a shared, public experience), collective homosexuality in South Africa is still in the embryonic stages of development in respect of public coming out. The reasons for this include:

▶ the lack of a unified, national homosexual liberation movement;

▶ the absence of legislation to counter discrimination;

- the general lack of support for homosexual identity and commitment;
- societal disapproval (including the fear of parental disapprobation), which tends to delay the coming out process; and
- the dearth of supportive organizations which could provide role modelling and care, including counselling units.

The above circumstances should be borne in mind when an examination is made of the extent to which coming out was noted as a crisis by study respondents.

'Coming out' crisis

Table 7.6 Experience of crisis in 'coming out'

Crisis	Respondents	Percentage
Yes	41	46
No	39	43
Uncertain	10	11
Totals	90	100

Forty-six per cent of respondents acknowledged coming out as a crisis. However, hardly any literature has described the actual phenomenon of crisis in specific regard to coming out, although many researchers have noted that coming out can be a debilitating experience.

Jay and Young (1979) record experiences of shame, guilt, poor self-esteem, self hatred, and isolation, while Daher (1977) and Isaacs (1977b) describe elements of extreme anxiety. Colgan (1987) shows how separation and attachment may delay identity resolution, commenting that coming out simultaneously evokes awareness of *difference* from other males, and awareness of *desire* for other males, thus creating dichotomous experiences of identity.

Other debilitating effects which observers report include internalized stigma, corresponding to the 'spoiled identity' syndrome; guilt; withdrawal from homosexual activities; depression leading to the need for psychiatric care; suicide and suicide attempts; and seeking ways and means to 'change' identity (Bell and Weinberg, 1978; Cramer and Roach, 1988; Gershmam, 1983; Jandt and Darsey, 1981; Muchmore and Hanson, 1982). Finally, Gershman (1983) and Colgan

(1987) both emphasize that irrespective of the situation, coming out will be stressful for the individual, and will place him at risk.

Further debate as to whether or not coming out entails the experience of crisis is raised by a letter received by one of the writers from the editors of the *Journal of Homosexuality*. Wendell Ricketts (the managing editor), on behalf of John de Cecco (the editor), writes:[3]

> Although integration of same-sex fantasies, relationships and behaviour into a single personality can be traumatic, not everyone assimilating these feelings experiences a crisis. Many people manage to accept their feelings quite calmly, either acting upon them or not, either maintaining other kinds of sexual relationships or not. People are utterly resourceful and ingenious in the ways they construct relationships and in the ways they include sexuality in or exclude it from these relationships.

This statement both denies the growth-promoting feature of crisis, and fails to recognize that crisis does not necessarily imply pathogenic behaviour, even if the experience is stressful or debilitating. Ricketts' contention that coming out is commonly free of crisis is contradicted by findings of the present research, which revealed that just less than half the respondents (46%) unambiguously identified a crisis, and that a further 11% were 'uncertain'. Of these latter respondents, five gave the following descriptions of their 'uncertainty':

▶ 'I was frequently afflicted by suicidal depressions, was very tense and uptight, and experienced feelings of alienation.'

▶ 'The crisis was extended over some years. Feelings of fear and anxiety were uniformly high for two years.'

▶ 'Not easy. Traumatic and problematic.'

▶ 'My mother had the crisis which affected me.'

▶ '... I recall a sense of loss. More in the realization that the reality of homosexual experiences did not correspond to my dreams/fantasies of what it would be like!!!'

Thus, a crisis situation certainly existed for these respondents, even though they had expressed doubt about it. It is therefore possible to state that coming out involved a crisis for more than half of the respondents. In order to preserve the qualitative responses of the respondents who unambiguously identified a crisis in their coming out, and who also made additional comments describing it (26 respondents in all), these comments are summarized in Chart 7.1.

Chart 7.1 Summary of comments made by respondents in respect of 'coming out' crises

Respondents (N=26)	Summarized commentary
1.	Psychiatric treatment, suicide attempts, rejection by parents.
2.	Engaged to be married, broke off engagement, nervous breakdown, resulting in hospitalization and shock therapy.
3.	Sense of aloneness, loving another man is sinful, fear of sex.
4.	Panic, fear, anxiety.
5.	Loss of self-esteem, lack of confidence, depression.
6.	Social anxiety, fear of parental rejection.
7.	Fear of heterosexual encounters and pressures.
8.	Fear of family or of friends finding out, hence feelings of shame and personal doubt.
9.	Avoidance of family and friends, self-imposed isolation, moving to another city to avoid being discovered.
10.	Constant state of panic.
11.	Fear of family and friends finding out.
12.	Poor self-esteem, relationship difficulty, poor day-to-day functioning.
13.	Depressed, guilty, with feelings of perversion.
14.	Suicidal, with one attempt at suicide.
15.	Isolation and loneliness.
16.	Social rejection, fear and anxiety.
17.	Anxiety followed by relief, but stayed in 'closet' for 30 years. Always felt that an impending crisis existed.
18.	Failed twice at university. Suffered severe depression.
19.	Saw distress in other gays. Identified with them, had a girl friend as well as a boy friend, felt fragmented.
20.	Crisis related to transvestite issues, confused: attempted suicide on two occasions.
21.	Family rejection, expelled from home.
22.	Self-doubt, anxiety, inability to articulate feelings for fear of being exposed as a 'moffie'.
23.	Parental disapproval, could not be open in public, felt like a fraud.
24.	Fear of rejection by family and friends, fear of loneliness and abandonment, 'thought I was the only person going through Hell'.
25.	Sense of complete disillusionment with the gay scene, self-imposed isolation.
26.	Feelings of inner torment, guilt, and sexual discomfort.

Two major features are apparent from the chart. First, many descriptions reflect crises of an existential or egocentric nature. All-pervasive feelings of inner despair, a sense of hopelessness, and indications of internalized homophobia are apparent. Second, the comments show that an attempt was sometimes made to resolve the crisis by acting out behaviourally. Examples include suicide attempts, self-imposed social isolation, estrangement from family and friends, the deliberate withholding of information, the dissolving of relationships, and geographical mobility.

This South African material shows a picture different from Ricketts' statement of the American experience. The American context might, however, well be different because of human rights legislation, and legislation (in certain states) protecting minority and sexual rights; the removal of homosexuality from the mental disorders list; strong gay activist and prominent gay rights lobbying movements; and a well-developed support system and network of gay alliances, including international affiliations. Conversely, the absence of such conditions in South Africa, and especially the lack of crisis services, might account for the accentuated sense of crisis experienced by many respondents on coming out. Supportive evidence for this statement is found when an examination is made of the respondents who found their crisis 'disturbing' (Table 7.7), and why they found it so.

Feelings about the crisis

Table 7.7 Experience of 'coming out' crisis as being disturbing

Disturbing	Respondents	Percentage
Yes	40	44
No	17	19
Uncertain	6	7
Unknown	27	30
Totals	90	100

Just under half of the respondents regarded their crisis as 'disturbing', 17 felt their crisis to be manageable, and six were uncertain. Twenty-seven respondents failed to indicate whether or not the crisis was disturbing, even though some of them had acknowledged having a crisis.

On the other hand, 12 respondents who claimed *not* to have had a crisis in conjunction with their coming out, responded by saying that they were disturbed. Their reasons included a threat of legal or police action, the leading of a double life with reference to employment, loss of employment through dismissal, and the need to father (or parent) children. Perhaps the word 'disturbing' in the questionnaire triggered a subliminal sense of discomfort, which elicited these responses. Of note, however, is that in contrast to those reporting an 'internal' crisis, or acknowledging a personal sense of distress, the 'disturbing' features in the above instances all related to broader dimensions involving public opinion or influence. In other words, the disturbing feature was not experienced as egocentric, but as *sociocentric*. This could be a defence against anxiety, since it is safer to acknowledge external or generalized issues, and thereby to compensate for inner discomfort.

Neither the time span nor the specific nature of the crisis associated with coming out was expressly asked for in the survey. Nevertheless, the extent of relative debilitation revealed by the survey, together with the fact that half of the respondents acknowledged a crisis, has considerable significance for the following reasons:

▶ coming out seems to be strongly related to crisis;

▶ the crisis responses of people in the survey generally had negative consequences for them;

▶ the crisis was not experienced as a single incident, but as an internalized set of responses within the context of a major identity struggle; and

▶ as Chart 7.1 shows, the crisis did not pass quickly and was often protracted.

Although for some respondents coming out represented a release of tension or a positive catharsis, Minton and McDonald (1984) concur that it is a significant event which can involve anxiety and confusion. These feelings are caused by a disparity between individual identity (self-perception) and social identity (presentation of self to others), and can be resolved through a process of psycho-cultural management. Since coming out is one feature of identity disclosure, the successful management of it is vital for the integration of homosexual identity into broader personal identity, thereby achieving a whole sense of self (Cass, 1984; De Cecco and Shively, 1984; Jandt and Darsey, 1981; Lee, 1977; Malyon, 1982b; McDonald, 1982; Weinberg, 1978).

How 'coming out' occurred

Table 7.8 How 'coming out' occurred

How coming out occurred	Respondents* (N=90)	Percentage*
'It just happened'	44	49
'It was forced upon you'	8	9
'You sought out other gays'	41	46
'You read about homosexuality'	34	38
'People talked to you'	22	24
'Through therapy'	11	12
'Some other way'	13	14

* Multiple answers are possible

Almost two-thirds of respondents attributed their coming out to two or more circumstances. This confirms the notion that coming out is rarely attributable to one factor alone (i.e. it does not suddenly 'happen overnight'). An accumulation of circumstances usually provokes 'coming out', so that it is a process and not a single event. The two most pronounced events in the process were that 'it just happened' (44 responses) and seeking out other gays (41 responses). Other research is in accord with this. Firstly, Richardson (1984) confirms that homosexuality 'happens' and develops from a state of being ('homosexual sensation') into sexual desire, sexual behaviour, and sexual identity. Secondly, the seeking out of other homosexuals represents an active searching for confirmation that is inextricably linked with the identity growth pattern. This behaviour has also been noted by other researchers, such as Hammersmith (1987), Hart (1984), and Hetrick and Martin (1987). The sub-culture is possibly the most powerful force in facilitating the coming out process, since it provides the arena for seeking out to take place. Although the respondents did not give much additional commentary in this regard, some indicated that the 'happening' occurred with a significant person who introduced them to other gay people. Additionally, it is of note that the 13 responses to 'some other way' in Table 7.8 all indicated that a special person (close friend/family member/ another gay man) helped to facilitate the coming out process. Other external factors which catalysed coming out included psychotherapy, homosexual literature, and talking with others.

A small number of respondents believed that their coming out had been 'forced upon' them. This calls for comment. Although these few responses came from respondents who indicated that they had been seduced or raped in childhood, such sexual exploitation cannot be regarded as a widespread, specific feature that determines homosexual identity (Jay and Young, 1979). The findings of this study contradict certain stereotypical myths which hold that homosexuality is caused through the 'corruption' of children by 'perverted' adults. Traumatic personal experiences in childhood must be seen only as possible predisposing factors, and must be placed within the entire repertoire of experiences (Daldin, 1988; Lee, 1977; Mieli, 1980; Sagarin, 1973; Weinberg, 1978). This point has relevance for the recent upsurge in public concern about, and police involvement in, cases of alleged child abuse and sexual molestation. Some public opinion has it that child molesters (or paedophiles) are homosexuals who were themselves seduced as children, and that they are thus perpetuating a pattern of learned behaviour. However, the findings of this survey suggest that childhood sexual abuse *per se* does not have a major bearing on homosexual identity.

The low response rate to psychotherapy as a force facilitating coming out must be seen in context. Although some respondents indicated that therapy increased their self-esteem and eased their confusion, their comments suggest that they did not see therapy as a major factor. Psychotherapeutic services are not always readily available in South Africa, and moreover some therapy practice has the reputation of being homophobic, since certain therapists have tended to advocate change into heterosexual patterns of behaviour.[4] The treatments reported by study respondents included aversion therapy, electric shock treatment, chemotherapy, and attempts at religious conversion. Some parents, upon discovering their sons' homosexuality, forced them to undergo therapy in order to 'change' them. It is only in the last few years, since the inception of GASA counselling centres in Johannesburg and Cape Town, that homosexual people have been able to receive appropriate and supportive therapy from gay organizations.

Other crises linked to homosexuality

As Table 7.9 reflects, more than three out of every five respondents (61%) indicated that they had experienced a crisis related to an aspect of their homosexuality other than coming out.

Table 7.9 Other crises related to homosexuality

Other crises	Respondents	Percentage
Yes	55	61
No	29	32
Uncertain	6	7
Totals	90	100

For the most part, respondents' commentaries confirmed that their 'other' crises were identity-specific, and essentially of a sociocentric nature. Minton and McDonald (1984) contend that sociocentricity emerges when the individual has a heightened awareness of possessing a homosexual identity, coupled with an awareness of societal attitudes concerning homosexuals. Other studies corroborate that the homosexual experience is strongly determined by society and social attitudes such as homophobia (Bohn, 1984; Ehrlich, 1981; Gramick, 1983; Morin and Garfinkle, 1981; Siegel, 1981). Weinberg (1978:143) advances the view that homosexuals reinterpret their behaviour as homosexual when (a) they perceive changes in the behaviour of their friends or significant others towards them, or (b) they come into contact with self-defined homosexuals.

The tendency to externalize the crisis can be linked to the notion of power (Focault, 1978). Experiencing crisis may be related to the absence of a central personal power base, particularly with reference to institutionalized masculine gender roles. Homosexuals often feel that they have to live up to the traditional images of masculinity dictated by the parent culture. These images or notions of masculinity are incorporated into homosexual imagery. But because homosexual imagery is usually judged by society as weak and effeminate, gays experience crisis as a result of the dissonance between external and internal symbols. Crisis may thus decrease a sense of masculinity, thereby perpetuating a sense of weakness. This point is exemplified in the following statement made by a client in therapy: 'I was afraid of coming out because I thought I would turn into an effeminate man.' Carrigan, Connell, and Lee, writing in the context of power and hegemony, support the above argument and believe that any kind of powerlessness among men readily becomes involved with the imagery of homosexuality (Carrigan et al., 1987:86).

Interestingly, no respondent in the study noted a crisis stimulated by religious guilt, yet religious factors have been well represented in the crisis profiles of a number of clients seen by the writers. In essence, religious doctrines, usually of the fundamentalist kind, have caused people considerable delay in negotiating their identity synthesis. Oppression of homosexuality is rooted in Judaeo-Christian traditions, which remain the firm foundation for socially accepted behaviour in Western cultures. The crisis of identity has strong connections with religious feelings, and can be the source of much distress. In the present research study, the absence of 'religious guilt' as an explicit cause of crisis related to homosexuality may have been due to respondents subsuming their religious difficulties under different notions, such as 'guilt', 'perversion', 'parental wrath', and 'conservative family background'.[5]

Overall, the findings of this study indicate that gay crisis comprises an egocentric stage *and* a sociocentric stage. This dual phenomenon compounds the process of identity synthesis and constitutes a key dimension in homosexual identity development. Crisis is not only an externalized response to an internal event; inner turmoil is also generated by perceived external factors including societal attitudes, 'normal' heterosexual functioning, religious doctrines, existing legislation, and so on. This often retards the experience of coming out.

Perceptions of coming out crisis in others

In addition to asking respondents directly about their own experiences of crisis, an opportunity was also given to them to comment — perhaps with more 'safety' — on whether or not other gay men have crises associated with coming out.

Table 7.10 Perceptions of coming out crisis in others known to respondents

Others have crises	Respondents	Percentage
Yes	81	90
No	5	6
Uncertain	4	4
Totals	90	100

An overwhelming majority of respondents (90%) indicated that they had known other gay people who had experienced a crisis in their coming out, and only six per cent said that they definitely had not. It is not possible to determine from this information the nature and intensity of the crisis in others, but respondents' answers provide a strong indication that crisis relating to coming out was, in their experience, applicable to members of the homosexual collective.

What respondents believe may indeed be true. Nevertheless, the point must be taken a step further. Perception of crises in other gays may be influenced by the myth that gays are crisis-prone. Hence, respondents' feelings about the 'gay crises' of others may be intensified as a result of antipathy towards the gay culture and gay relationships in general. Many gay people anticipate gay crises without necessarily experiencing them — this is part of the gay homophobic syndrome (Ehrlich, 1981). It follows that a biased anticipatory fear may be created in respect of feelings of intimacy, and in consequence inner feelings of hostility are projected on to the gay sub-culture. This internal distance from gay objects, related to the terrifying consequences of being labelled 'gay' or 'homosexual', allows for the projection of crisis on to a collective persona i.e. other gays.

Perceived differences between homosexuals and other people

Table 7.11 Perception of difference between homosexuals and other people

There is a difference	Respondents	Percentage
Yes	53	59
No	25	28
Uncertain	12	13
Totals	90	100

Nearly three-fifths of the respondents considered that gay people in general believe that they are significantly different from 'the man in the street'. This suggests a notion of homosexual separateness. As previously indicated, the gay sub-culture perpetuates a sense of 'difference' or 'specialness', primarily in order to nurture the gay ethos. From a more pragmatic point of view, this sense of difference has a historical legacy, for medical and socio-legal assertions have traditionally typecast homosexuals as 'different'. Part of this difference has emerged from labels, such as *disordered, deviant, promiscuous, queer, perverted*, and the like. Homosexuals have thus had good reason to think of themselves as separate and distinct. As a result, the gay sub-culture has gained its recognizable modern form by pronouncing this very sense of difference, and promoting it via its iconography, symbols, vernacular, behavioural styles, and sexual expression.

The sub-culture also serves as an extended surrogate family for the gay collective. In fact, gay people refer to other gays in their social orbit as 'the family'.[6] This creates an immediate sense of belonging, which simultaneously rates the person as different from others by virtue of being gay, and embraces him into the symbolic family network.

Perhaps the most constant feature in a homosexual person's life space is the institutionalized ritual of the sub-culture, which has a universal set of codes and behaviours yet, like a family system, also maintains and perpetuates its own idiosyncratic style. Hartman (1978) speaks of the eco-map of a family, which reflects a family's structure, its systems, its extended network of interactions, and the sources of nurturance, stimulation, and support that are essential for the survival of the family. This ecological metaphor is of assistance in understanding the notion of the 'gay family', which may be broken down into two principal components, namely the micro-family and the macro-family.

The micro-family fosters the sense of difference according to local (indigenous) factors. It comprises separate groupings of people, such as English- and Afrikaans-speakers, the 'sophisticated set', the 'political alternatives', and the drag queen set, which at times collectively participate in the wider sub-culture. The micro-family thus tends to divide and compartmentalize gays into cliques, by emphasizing differences of maturity, lifestyle behaviour, values, education, politics, language, and culture. In so doing, it compounds the process of identity development by perpetuating the experiences of duality and tripartite confusion, discussed earlier in Chapter 1.

The macro-component universalizes the gay context, and creates the foundation for the rules and regulations which govern gay behaviour. These rules, according to the symbolic interactional paradigm, filter down to the micro-system, causing various alignments and power coalitions. Minuchin (1974) uses the concept of 'enmeshment' to describe the interaction which characterizes such sub-systems. Enmeshment refers to 'being trapped in the biography of the family', with the behaviours of members creating an immediate effect on each other.

Enmeshment, in the context of 'gay difference', refers to

▶ The need to perceive and experience the sub-culture as separate and distinct from other cultures, including the parent culture.

▶ The need to create alternative family systems which respond as surrogate caretakers.

▶ The need to perpetuate a sense of difference by ensuring that the ritual of the sub-culture remains within the surrogate family. (Should there be a fusing of the sub-culture (gay family) with the parent culture (heterosexual society), the identity base would become blurred, thus disturbing the homeostasis of identity needs.) However, only when appropriate disengagement from the sub-culture occurs are people likely to perceive themselves as distinct individuals, thereby acquiring the status of 'field independent' (see Chapter 1). This feature is noticeable in certain comments offered by some of the minority of respondents who believed that homosexuals (or gays) were not significantly different from others.

In determining the fundamental nature of the difference asserted by the majority of respondents, and noting the justifications which they offered in support of their judgement, the following may be suggested:

▶ The difference is primarily sexual.

▶ The difference is based on a collective notion of 'sensitivity', or an undifferentiated gay ego, which upholds the universal myth that gays are more sensitive, artistic, emotional, and vulnerable than non-gays.

▶ The difference is perpetuated by the 'ghetto mentality' or 'cloning syndrome'. Some respondents indicated that they see themselves as the gay archetype first, and then as people. De Cecco believes this sort of thinking to be a gay socialization process, which promotes the image of the gay individual as a distinct identity, particularly in order to gain minority group status (De Cecco, 1984b).

▶ The difference is maintained because of victimization and varying degrees of oppression.

A more recent view of the notion of difference has been expounded by Carrigan, Connell, and Lee. They suggest that the social space of homosexual relationships is rapidly changing, and that these relationships are now less marked by the rules of gender division that are dominant in heterosexual society. Distinctions between the 'active or passive' or 'masculine or feminine' homosexual man may have lost their former significance (Carrigan et al., 1987:88).

Attitudes towards a service centre for people with homosexual concerns

Table 7.12 Attitudes towards a service centre

Would like a centre	Respondents	Percentage
Yes	79	88
No	3	3
Uncertain	8	9
Totals	90	100

The great majority of respondents (88%) thought that a centre should be established to deal specifically with homosexual concerns. This high response rate was based on three aspirations which were revealed in respondents' commentaries:

▶ A centre could serve as a legitimate unit of interaction between gay people. It would possess the status of a bona fide organization which would not be directly associated with a gay venue that ascribes to the cult of sexuality.

▶ The centre could function as an arena for 'becoming public'.

▶ The centre could operate as an independent service organization which would be homophilic rather than homophobic.

Historically, gays in South Africa have had to rely on agencies that are predominantly heterosexual, and private practitioners (not necessarily gay or gay-sympathetic) for assistance. Research in this respect, notably that of Dardick and Grady (1980), De Crescenzo (1984), Greenberg (1976), Margo (1976), and Messing et al. (1984), consistently affirms that many health care professionals display homophobic responses in their dealings with homosexuals.

However, helping organizations established by homosexuals for homosexuals are not without problems. For instance, Greenberg (1976) reports as follows on the effects which a homophile organization has on both the self-esteem and alienation of its members:

> The data indicated that new members [to the Mattachine Society, a homophile organization] could expect to feel a greater control

over their destiny – decreased powerlessness – and an increased sense of having rules, regulations and standards with which they could abide – decreased normlessness (Greenberg, 1976:316).

The generalizability of Greenberg's study findings should be treated with caution, because of his limited sample size. Nevertheless, homophile organizations (including counselling centres) are undoubtedly safe and understanding helping resources for gay people.

The homophile organizations which exist in the United States have a strong activist, lobbying power base, and are inextricably linked with liberation attempts (Rueda, 1982). By contrast, most respondents in the present study agreed with the 'softer' option of a helping centre, which would presumably deal with individual needs rather than addressing social challenges through collective action. On the other hand, the three 'no' and eight 'uncertain' responses reflected in Table 7.12 indicate that some of the respondents were unhappy with a purely service-orientated organization. Their additional comments revealed a belief that such an organization would promote the 'predicament' or 'plight' of the homosexual, and therefore alienate him further from the mainstream culture.

Nevertheless, the results of the present study affirm that most respondents wanted a centre providing help for people with homosexual concerns. Their wishes were subsequently fulfilled with the advent of a counselling service which grew parallel with the establishment of a community centre, known as the GASA 60-10 Community Centre (Pegge, 1988b).

Perceptions of, and attitudes towards GASA 60-10

Table 7.13 Perceptions of, and attitudes towards GASA 60-10

Perception/attitude	Respondents* (N=90)	Percentage*
Could help with coming out	42	47
Alleviates loneliness	27	30
Facilitates meeting people	60	67
Bores me	12	13
Dislike it because it caters for gays only	4	4
Members are snobbish	6	7
Could provide relevant information for gays	76	84
Other	9	10

*Multiple responses are possible.

Table 7.13 shows positive responses to eight statements about GASA 60-10. The item with the strongest response (that the organization could provide relevant information for gays) underscores respondents' wish for a service centre, noted earlier. The second strongest response (67%) was that the organization facilitated meeting people. Hence, it is surprising that the apparent converse of this (alleviating loneliness) had a response rate of only 30%. Loneliness is a widespread psycho-social condition that confronts homosexuals. Besides the fact that the majority of clients seen professionally by the writers indicated that loneliness was an all-embracing feature of their lifestyle, figures from the *GASA 60-10 Counselling Annual Report* confirm this (Pegge, 1988b:1–2). However, respondents to the questionnaire might have seen the need for 'meeting people' to be of the same order as 'alleviating loneliness'. Alternatively, 'meeting people' might have a sexual connotation, while 'alleviating loneliness' might not.

It is important to note that GASA 60-10 was perceived as a vehicle that could help in the coming out of members (47%). In South Africa there is still a continuing private and public reticence about homosexual identity, and a fear of being perceived to be homosexual.

In the United States, Lee (1977) argues strongly for recognition of the concept of 'power' in the coming out period. Power comes from belonging to a 'gay group', which contributes to the person's ability to control his situation and his future. Lee itemizes the North American coming out process as follows:

▶ *Step 1*: first debut;
▶ *Step 2*: regular at bars;
▶ *Step 3*: coming out to heterosexual friends;
▶ *Step 4*: coming out at work; and
▶ *Step 5*: in a gay liberation group.

Step 5, the last stage in Lee's graduation process, is linked to power. Lee states: 'The act of going public, while seemingly an act of powerful individualism, is in one sense a rejection of the individualist ideology in favour of the concept of community' (Lee, 1977:64). However, there is little sense of a collective community in South Africa, and hence it is safer for gay people to assign a 'helping' status to a local gay group. Helping others to come out is a far more acceptable goal than 'liberation politics' for a group of people who are themselves still struggling with the concept of going public, and who have not yet reached the stage of identity synthesis (Cass, 1979, 1984; De Monteflores and Schultz, 1978; McDonald, 1982). Going public, even within the context of a gay group, involves the recognition and acceptance of *community* (Hodges and Hutter, 1974). This has not yet been achieved in the South African gay context. Simply put, gays do not trust their 'families' (i.e. the micro- and macro-families discussed earlier), even though they need them. This confirms the 'double bind' status of homosexuals described in previous chapters.

An important feature of the data reflected in Table 7.13 is the strikingly low positive response rate to the items designed to tap possible gay homophobia ('it bores me'; 'caters for gays only'; and 'members are too snobbish'). A possible explanation is that participation in the GASA 60-10 group was still a relatively novel experience, since the group had not been operational for many years at the time of the study. However, during the subsequent maturation process of GASA 60-10, it has been noted by one of the writers that gay homophobic features have indeed become operational. Members have left for the reason that the group is gay only, its members are snobbish (a euphemism for 'uptight queens'), and that they are bored. This notion of 'boredom' needs explanation. Boredom in this context relates to the lack of homo-erotic stimulation. A disillusion occurs, including a substantial sense of disappointment, about the absence of erotic features such as freely available opportunities for sex, a fixation at sexual levels of camping, and overt sexual behaviours. One respondent stated in his comment: 'I'm tired of the same old faces. I want excitement. It's like coming home to a boring family. I'm sick of the closed and incestuous nature of 60-10.'

The paradox reflected by the foregoing findings and commentary must now be addressed. While the respondents indicated that GASA 60-10 would facilitate meeting people, the gradual lack of interest displayed by members (epitomized in the commentary cited above) seems to be closely related to the notion of impersonal sex, or sex without commitment, obligation, or a long-term social relationship. The emphasis is on the transiency of sexual experiences and the homo-erotic. While GASA 60-10 was initially perceived as a place where social relationships could be developed without the accent on sexuality, over time pervasive disappointment became evident, corroborating the remarks made by Weinberg and Williams (1975) in their benchmark article on 'Gay Baths and the Organisation of Impersonal Sex'. They write:

Because of the singular purpose often involved in impersonal sex, many males do not want a complex or broad social relationship. Thus from their perspective it is desirable to limit non-sexual social interaction. This desire is often related to the wish to conceal the activity or to avoid involvements that could compete with established relationships (e.g. romantic relationships). It is also related to shyness or a wish psychologically to compartmentalise the activity. In effect, none of the aspects of a primary relationship would appear. The interchange would be easily transferable from one partner to another and narrowly confined in its social depth and breadth (Weinberg and Williams, 1975:131).

Weinberg and Williams' observations, along with the findings of the present study and the writers' additional commentary, lend support to the critical significance of the transitional object as a factor in homosexual identity development.

Desire to help gay people in distress

Table 7.14 Desire to help gay people in distress

Wish to help	Respondents	Percentage
Yes	70	78
No	9	10
Uncertain	11	12
Totals	90	100

Although the findings contained in Table 7.14 might be interpreted as an altruistic response, it can be suggested that further psychological variables might also be evident. For instance, by dealing with someone else's vulnerabilities, perspective may be gained of similar problems or positions experienced by the respondents themselves. Furthermore, helping others is a safe (or relatively safe) way to deal with one's own coming out status. Measures of 'worse' or 'better' can effectively be matched against the predicament of others. This could either detract from the person's own experiences, or add to an already existing burden. A measure of 'gay curiosity' or 'gay voyeurism' can be present, and comparative situations can be explored and dealt with through shared victim-like responses. In this case, altruism would reflect a negative identity, which would be reinforced by the mutual and reciprocal experiences of the counsellor and the counselled.

Conversely, however, Smith (1988) warns that altruistic responses are not necessarily always linked to avoidance of anxiety and maladaptive responses. Many homosexually adjusted people who have dealt with (a) loss, (b) the burden of homophobia, and (c) social stigmatization have developed 'superior coping skills' and, within the context of a consolidated self-image, are in a position to be helpful to other people (Smith, 1988:70).

Preferred ways of helping

Table 7.15 Preferred ways of helping

Preferred way	Respondents* (N=90)	Percentage*
'on your own'	43	48
'via the 60-10 group'	36	40
'through some other formal organization'	29	32
'through some other channel'	17	19
'uncertain'	22	24

* Multiple responses are possible

Slightly more than half of the respondents (54%) identified only one preferred way of helping gay people in distress, while the remainder identified two or more ways, or were uncertain. The options of helping 'on your own' and 'via the 60-10 group' were the most popular.

Respondents who indicated that they would prefer to help people on their own were drawn largely from the category of highly skilled people described at the beginning of this chapter, and included social workers, doctors, lawyers, and psychiatrists, who suggested that their skills would enable them to handle gay issues at an individual (and private) level.

It can be suggested that this finding is capable of symbolic interpretation as well. It will be recalled that Table 7.8, which reflected feelings about coming out, included the item 'it just happened'. This 'happening' was often facilitated by a 'significant other' who helped with the coming out process. This coincides strongly with object-relations theory, which deals with the learning cues specifically associated with sex role identity during infancy (Murray, 1968).

Coming out may be described as a 'rebirthing experience', and the association with a significant person during this process is imprinted in the person's psyche. Coming out in most cases induces a state of psycho-social vulnerability, and psychological regression is not uncommon. In crisis theory terms, this renders the person extremely vulnerable, with minimal defences, and highly susceptible to external influences. Therefore significant people replace the biological models (parents) in relevance and importance. This, too, may be the prelude to the initiation of the gay 'micro-family' as discussed previously. It has not been uncommon for the writers, during many professional experiences of facilitating the coming out process for individual clients, to be referred to as 'the midwife' (the person assisting the birth process). This metaphor emphasizes the symbolic importance of the significant other in contributing to the discovery of identity. It was perhaps in recognition of this that some respondents felt that they would like to help others on their own.

The GASA 60-10 group drew the second highest response rate (40%) as a means for helping gays in distress. This is an expected outcome, for the respondents in the main perceived or believed that their organization could fulfil a direct service function.

Commentaries offered in response to the other ways of helping shown in Table 7.15 came chiefly from people already bonded to service organizations such as Life Line, welfare organizations, and religious groups, in other words from social workers, priests, pastoral workers, personnel officers, and the like. Respondents who reflected uncertainty in the means of offering help to others were generally vague, but one respondent noted that his uncertainty was prompted by the 'semi-confidential' nature of counselling, and added: 'Gays have big mouths and are bitches and will spill the beans to all and sundry.'

Feelings concerning being a research subject

The final question in the questionnaire sought to ascertain whether or not respondents felt that their privacy had been invaded by being asked to participate in the study. The question was prompted by multiple considerations:

▶ A suspicion that systematic research into the private world of gay people would reinforce their status as research-worthy because they are 'different'.

▶ Some respondents in the pilot study were uncomfortable about confidentiality and the purpose of the research.

▶ A belief that the question could elicit angry or passive-aggressive responses, as well as certain attitudes towards scientific research into homosexuality.

▶ An opportunity would be provided for the respondents to be 'debriefed' after filling in the data (this could provide an outlet for negative and positive feelings and for commenting on their own sense of participation).

Table 7.16 Feelings of infringement of privacy by the study

Infringement	Respondents	Percentage
No	83	92
Yes	5	6
Uncertain	2	2
Totals	90	100

Nearly all the respondents indicated that they did not regard the questionnaire as an invasion of their privacy. Their comments were warm, supportive, and congratulatory. Many indicated relief that research was being undertaken. Others noted that the research could ultimately provide apposite information for both the homosexual and the heterosexual populations. Respondents' positive comments (e.g. a call for the 'defusing of myths', 'it's about time research is undertaken', and 'maybe we can learn about ourselves') coincided with the theme of 'needing relevant information', discussed previously.

The minority of respondents who felt that the research study was an infringement of their privacy (6%) all indicated a fear of betrayal in respect of personal confidentiality. It is assumed that either they had had uncomfortable experiences with breach of confidentiality in the past, or that they believed their responses could expose them 'publicly'. The two respondents who were uncertain offered no explanatory comment.

Although the tenor of responses to the present study indicated a considerable show of support, this is not always the case. A recent research study undertaken by a post-graduate student at a university in the Transvaal was slated by the national gay publication, *Exit*. In an editorial, the gay collective was cautioned not to co-operate because the credentials of the researcher were suspect (*Exit*, 29,

June/July 1988:1). In the present study, close and open collaboration with GASA 60-10 in planning and implementing the research was one of the features which contributed to the unusually high degree of participation by potential respondents.

The findings of the study reported upon above support many of the postulations which the study sought to test. In the following chapter, conclusions are drawn based upon study findings and the theory perspectives provided in preceding chapters.

Notes

1. It must be remembered that the establishment of GASA 60-10 was networked primarily from the gay populace in the Greater Cape Town area.
2. Personal communication with senior flight attendants of South African Airways. This has also been confirmed by Mr J Pegge of the AIDS Support and Education Trust in Cape Town.
3. This was a response to a request by one of the writers to the editor of this journal, where additional information pertaining to published material on crisis, coming out, and homosexuality was asked for.
4. Some of the writers' clients have related severe homophobic responses from previous therapists. Part of the homophobia is reflected in the treatment methods used, which have included shock and aversive therapy. However, attention is also drawn to responses from some therapists who have intimated to their clients that homosexuality is 'fine' and that no intervention is necessary. This is a response that is often at variance with what the client *feels*, and which induces a panic state since the reality of the client and that of the therapist differs considerably. This point should be taken into consideration by helping professionals when dealing with coming out crises. No assurances should be given to a client unless both people are aware of the internal and external realities that confront them.
5. This is a limitation of the questionnaire. It did not specifically address religious factors within the context of crisis. Had religion been separately categorized in order to ascertain the link between religious feelings, crisis, and identity, responses in this regard might have been different. Unfortunately, religion was omitted from the questionnaire at the request of the GASA 60-10 Committee, who believed that issues of race, ethnicity, and religion were too delicate to be addressed at the time.

6. The notion of the 'gay family' has been institutionalized in a world pop hit by a gay Australian music group, The Village People. The disco song 'We're Just Family' universalizes the notion of the extended homosexual family.

8 Conclusions and implications for helping professionals

Human sexuality, including homosexuality, forms an integral part of the many human concerns dealt with by helping professionals, such as social workers, psychologists, medical doctors, and psychiatrists. Hence, a helping professional cannot avoid sexuality issues without negating values that are part of his or her profession's ethical foundation, such as respect for basic human rights and dignity, and freedom of choice.

The present chapter seeks to promote increasingly appropriate engagement between helping professionals and clients with homosexual concerns by presenting the main conclusions which can be drawn from this book, and then examining the implications which this knowledge has for effective professional intervention.

Conclusions

1. Homosexual identity development is complex and ongoing

The research study reported upon in Chapter 7 confirms the notion that the formation of homosexual (or gay) identity is a complex and ongoing process. The homosexual sensation, or 'happening', is not sudden; it occurs as a gradual unfolding experience, with no suggestion of genetic or hormonal factors. The child negotiates a private notion of sexuality within the parameters of his family, and while family experiences may contribute to the development of a homosexual identity, they do not 'cause' it.

In the broadest sense, a child's basic need for succour is contingent upon both dependency and attachment needs. According to Teyber, if the parenting figures are consistently responsive to the child's bid for affection, the child will freely be able to experience and express this need. If the parent is not responsive, or undermines the child's sense of expression, which might include the acting out of sexual fantasies as described in Chapter 1, then anxiety and its associated feature of poor self-esteem will soon become associated with the child's need (Teyber, 1988:115). In this way, a dichotomy may be created between self-identity based on an internal fantasy structure, and the social environment, which may be a precipitant for crisis in the expression of homosexual behaviour.

2. Identity development is influenced by the presence or absence of crisis

Findings of the research study support the view that gay identity is influenced by crisis. Although little empirical evidence for this notion is to be found in the existing literature, the established causal antecedents of identity crisis primarily concern ambivalence in the relationship that the child/adolescent has with his parents (Baumeister et al., 1985). When crisis theory and intervention are related to the development of homosexual identity, the following observations may be made:

▶ Human developmental patterns can be described according to crisis protocols. These include the crises of transitions, developmental crises, situational crises, and role crises. All crises, by definition, are accompanied by *loss*, and associated potential for *gain*. Crisis resolution is the successful negotiation of events in the life transition of the individual. By implication, loss must be dealt with before gain can be achieved. Because consistent loss factors (including the loss of heterosexuality) bombard homosexuals, and because they are denied the expression of the appropriate mourning, grief, and anger which are natural components of the loss situation, they often incur a fixation at the loss level, with gains delayed until the loss can be dealt with.

▶ Crises may be expressed or non-expressed, anticipated or unanticipated. The more familiar (and acceptable) the crisis, the easier it is to negotiate. As noted in the research study findings, *socio-*

centric features (i.e. responses accrued from external social pressures) are relatively easily identified and dealt with. Dealing with the external blame perceived and experienced by gay people may, however, promote a false sense of comfort, for the *egocentric* components of crisis are not located in the wider parent culture, and their resolution is considerably more complex and difficult.

▶ Crises may be normal or pathogenic. Within the homosexual context, because the expression of crisis may be private or camouflaged, the sense of appropriateness and normality is minimized or negated. The private fantasies of sexual attraction and the subsequent acting out of them (often occurring during puberty and early adolescence) are not publicly sanctioned. The behaviour and its associated crisis profile includes desperation, urgency, impasse, helplessness, fear, anger, and panic. This perpetuates a sense of weakness and loss of control. Crisis is thus linked to danger, and the associated features of gain, challenge, and hope are consistently underestimated. Therefore self-esteem and feelings of internal worth are often at risk.

3. Identity development is cumulative, with cerebral and behavioural components, and unsuccessful negotiation of stages of identity reinstates past conflicts, which may accelerate or retard the identity process

The research study findings indicate that homosexual identity takes place via a series of cumulative events, which have cerebral and behavioural components. As early as infancy, a repertoire of fantasies and experiences contributes to the development of identity. Internal fantasies, often supporting homo-erotic images in combination with same-gender attachment, are precursors to homosexual identity. Early experiences in childhood are stored in the memory, and are easily brought to the surface, becoming vital contributory factors to the growth of later homosexual identity patterns.

The research findings support the notion that the identity growth process carries residual experiences within it. This feature of 'unfinished business' presents in later years, and is often triggered off by a crisis. The stages of identity development outlined in Chapter 1 deal with the hierarchical experiences of identity acquisition; unresolved items at earlier stages of the identity agenda carry the possibility of unfinished business into later stages. For example, this is manifest in the AIDS crisis. Because AIDS is in most cases linked to sexuality, people have been forced to rediscover and renegotiate their unfinished business. Coming out issues, multiple sexual experiences, confronting the sub-culture, and being forced to disclose intimate experiences to people in a usually hostile parent culture, all contribute to the resurfacing of suppressed identity issues. Hence, even those who think that their identity has been resolved can experience a breakdown in their homeostasis.

4. Coming out is not synonymous with homosexual identity

The often-used term 'coming out' is not synonymous with the acquisition of a gay identity. Coming out, as the research findings have shown, is only one feature in the identity growth process. Coming out implies experiences or events which contribute to the individual's belief that he is gay. However, it can be argued that coming out is a life-long task which does not pertain solely to the traditional notion that sexual intimacy between men warrants a gay identity. Coming out may involve non-sexual features, such as openly declaring an allegiance to gay liberation, and coming out to employers, parents, and others. Thus, coming out is the accrual of homosexual experiences culminating in a synthesis of meanings and events.

Throughout this book, including the evaluation of the findings of the research study, coming out is marked as a significant process for the individual, with an attached set of meanings. Although coming out usually occurs during mid- to later adolescence, it has been noted that this process may be delayed until subsequent (adult) years because coming out and 'closet behaviour' are often linked. These two phenomena cannot be viewed as mutually exclusive. For example, some homosexuals may engage in same-gender activities without the public or significant others being aware of such behaviour, which may account for the number of homosexuals who are married, or who claim to be bisexual. Such people may have acknowledged their homosexual identity to themselves, but have withheld it from others. They have thus taken the first step towards coming out, but still remain in the closet.

5. Homosexual identity development begins at an early age, becomes definable at puberty, and takes its course within the parameters of the sub-culture

Research study findings provide strong support for the notion that homosexual identity growth is a developmental pattern with distinct features. Homosexuality has its genesis in early years, and recall experiences are noted at around five to seven years. Some notable features include aspects of cross-dressing, and existential feelings of being apart or separate from the family. The homosexual sensation (or experience) often occurs in this period. Recent evidence contradicts the premise that dysfunctional family units or absent fathers or dominant mothers produce homosexual sons (Clarke, 1977; Colgan, 1987; Cramer and Roach, 1988; Hetrick and Martin, 1987). Indeed, some research respondents, in the course of their comments, indicated strong bonding and caring relationships with both parents, and one in particular reflected on the warm and loving relationship that he had with his father.

This study confirms the belief that the early periods of homosexual sensation take on a different set of meanings at puberty, when the cerebral component is translated into sexual awareness. Puberty and its features of sexual development, arousal, and cognitive interpretations, fuel the homosexual sensation. Late adolescence is the critical period for homosexual identity. Adolescence is underpinned by crisis issues, and compounding homosexual fantasies (or instances of same-gender experimentation) serve to reinforce the inherent fear of being 'different'. Although respondents in the present research study demonstrated a mean age of 22 years as the significant identifiable time of coming out, the age spread for this is from early adolescence right up to middle age (i.e. 50 years of age and older).

The meaning of homosexuality and its associated experiences only becomes ordained within the parameters of the sub-culture. The sub-culture (in most countries where legitimate expression is allowed) consists mainly of young adults. In South Africa, coming out, which usually begins in mid-adolescence, is stalled until the individual reaches the age when the sub-culture can legally accommodate him. This hiatus between the time of inner experiences and the time when outer validation is possible, promotes dissonance in the growth process and contributes towards crisis. The psychological vulnerability which precedes sub-cultural access is usually associated with denial behaviour, manifested in attempts to conform to heterosexual role model expectations, which disguise the real features of identity structure.

6. The homosexual sub-culture produces a double bind for the gay person

The findings of the research study indicate a specific need in respondents to belong to a homophile organization, with apparent emphasis on the opportunity this provides to express their individual altruistic tendencies. However, on the basis of extensive findings reported in the literature as well as the clinical experience of the writers, there is no doubt that the sub-culture acts as a catalyst to the identity process. The sub-culture's emphasis on creating a gay ethos within the parameters of sexuality reinforces the notion that the sex components of homosexuality have precedence over and above all other intimate expressions of behaviour. The sub-culture provides a variety of role models, ranging from stereotypes (sex workers, bisexuals, transvestites, transsexuals, and 'butch'-'femme' types) to those who display no marked deviation from conventional descriptions of heterosexuals. The emphasis on role models facilitates sexual liberation, but it also stunts emotional actualization because of a fixation at the sexual level.

The importance of the sub-culture in facilitating the process of identity cannot therefore be underestimated. The range of options which the sub-culture opens to the gay initiate is overwhelming, often creating a sensory overload which tends to reinforce the transiency of sexual experiences. In turn, this endorses the primacy of the sexual component of identity. The sub-cultural tradition of affirmation through sex becomes lodged in the repertoire of the homosexual persona. This creates the notion of promiscuity which has beleaguered homosexuals. Promiscuity, described in Chapter 3, has bearing on identity in so far as it reinforces the self-worth of the individual. Sexual affirmation serves to compensate for the sense of separateness and internal isolation experienced during the early years.

Ultimately a homosexual person becomes gay because of acceptance into the sexual arena of the gay sub-culture. Sexual reinforcement validates the gay identity, and a relinquishing of this aspect renders the person alien to the culture, and liable to exclusion. This double bind entices the individual into the culture of sexuality. However, once the individual perceives and experiences the sexual transiency, a sense of despair unfolds, and the sub-culture is 'blamed' for promoting the *status quo*. If, for whatever reason, the individual rejects the sub-culture, he then faces the possibility of isolation, estrangement, and anomie. With no access to homosexual support in the parent culture, in addition to the disadvantages experienced in the sub-culture, the identity structure of the individual is bruised. To complicate matters further, as a result of reduced self-esteem, loneliness, and alienation, possible renegotiation of access to the sub-culture takes place. This often elicits passive-aggressive responses, stimulating geographical mobility and frantic searching for the idealized partner. The accrual of such experiences often leads to further antipathetic attitudes towards the sub-culture, and internalized homophobia is incorporated into the structure of the personality.

In consequence of the above, attention must be drawn to the notion of 'sexual addiction', and its clear distinction from 'promiscuity'. Sexual addiction, like any other addiction, denotes dependence on a substance or a situation (e.g. gambling) which takes on a set of compulsive behaviours. Immediate satisfaction of needs, over and above other factors, becomes the priority. Homosexual sexual addicts, like any other human addicts, fall into a separate and distinct category of people.

The notion of 'promiscuity' is a moral one (Altman, 1986), and does not take into account cultural and sub-cultural traditions and mores of wider populations. Indeed, society has institutionalized certain forms of promiscuity within the heterosexual context. For example, serial monogamy (i.e. a series of marriages or monogamous partners), Muslim traditional polygamy (i.e. more than one wife), and black concubinage might not incur opprobrium in specific communities (Moodie, 1988; Mokhobo, 1988; Van Onselen, 1982).

Promiscuity in the gay sub-culture is partly associated with obtaining an affirmation of identity from significant others. It can be a prerequisite for healthy endorsement of an identity that did not exist during the formative years and hence was never given an opportunity for legitimate expression. Sexual addiction, on the other hand, needs to be addressed as a separate and distinct feature, and dealt with as an addiction. The implications of this distinction are vital, and helping professionals need to take cognizance of the difference.

Finally, the homosexual sub-culture in South Africa maintains its self-credibility through the promotion of sexual interaction. The primary features of the sub-culture are institutionalized in the tradition of 'camping' and meeting people within a sexual imagery network. Discotheques, steam baths, bars, and camping locations are the temples of the sub-culture. Attempts to broaden the sub-cultural world into other areas are manifest in South Africa, but take second place to those already mentioned. Gay political groups, cultural groups, religious groups, and sports groups have tried to defuse the sexual component, but with little success, and these organizations occupy only a minor position in the overall sub-cultural spectrum.

7. AIDS has a direct bearing on homosexual identity development, and exists in a context of crisis

The advent of HIV infection and AIDS and the idea that it is a 'gay plague' has resurrected a wave of anti-gay feeling that is still washing over South Africa. The re-emergence of stigma, spoiled identity, and retribution forms the common base for internalized scapegoating. Gay identity issues which are influenced by society are profoundly affected by the AIDS scare. Two trends are discernable: the negative impact on homosexual identity, and the growth-promoting features of AIDS.

Negative impact on homosexual identity

Because of the direct association of AIDS with homosexuality, gay people see it as a threat to their existence. The transmission of the virus occurs mainly through blood and semen, and the primary means of spread is sexual, particularly via anal sex. The symbolic meaning of sexuality has now become contaminated. Because of the powerful association of the homosexual sub-culture with sexuality, the sub-culture is perceived to be contaminated as well. This has created panic within the gay collective, whose members believe that their form of metaphorical procreation has been jeopardized. The implications for coming out, including the process of identity synthesis, are great. Young people are unable to experience a fluidity of sexual experimentation (a prerequisite for aspects of identity growth) without fear of contamination and ultimate death. Homosexuality has thus acquired a *Thanatos* link, and coming out in the gay world is regarded with suspicion and fear. Retribution and blame now form part of the psychological pattern of dealing with the issue of coming out. Homosexuals who need to negotiate their identity via the sexuality route are now confronted with a meta-crisis situation:

▶ They have to deal with a deeply suspicious sub-culture.

▶ All forms of sexual intimacy have to be negotiated by safer-sex practices, yet these remain threatening for the inexperienced.

- Safer sex might deny the fulfilment of certain ingrained sexual fantasies. Consequent disappointment and resentment may lead to the quick dissolution of relationships, once again strengthening the transitional object syndrome.

- A psychological paradox of frightening consequences occurs when, during crisis intervention, some gay people admit to the hope of acquiring AIDS, so that they do not have to deal with the pain of coming out. In such instances, the *Thanatos* issue, or death wish, plays havoc with their identity.

- Levels of sexual experience and prowess are sometimes highly differentiated among homosexuals, for example, active/passive style, fantasy, and the like. The threat of AIDS demands the observance of safer-sex practices which may produce a sense of unfamiliarity, disappointment, and panic, which in turn can convert into performance anxiety. Hence professional attention is often needed to address sexual dysfunction.

- Patterns of denial still exist, even among the informed, and although the education drive within the sub-cultural context has gained rapid momentum, AIDS is perceived as an affliction that will strike others.

On the other hand, the AIDS crisis, both individually and collectively, has presented strong potential for crisis resolution in that it also embraces the growth-promoting features of crisis, such as danger, opportunity, and hope:

- AIDS has coalesced the splintered gay collective with a common aim: to defuse the myth that AIDS is a 'homosexual disease', and to educate its community. This has unified the gay community in one respect. However, as overseas experience has demonstrated, the fixation of AIDS concern on homosexuality must be broken, and broadened into the whole wide field of human sexuality.

- AIDS has allowed people to renegotiate their lifestyles as a result of evaluating their positions as risk people. Patterns of intimacy that were once taken for granted (such as those described in Chapters 3 and 5) are acquiring different priorities and emphases.

- Health hazards and lifestyle patterns are being prioritized according to need. AIDS has dramatically brought homosexual sex patterns out of the closet. Taboo subjects are discussed openly. Furthermore, health authorities, despite being accused of a patronizing attitude, have drawn attention to the world (and the plight) of the homosexual.

8. Alternative or fringe sub-cultures impact upon the identity struggle of gay people

Chapters 3 and 4 highlighted the territorial aspects of the gay sub-culture. The rise of fringe or 'alternative' sub-cultures, which ostensibly challenge fundamental traditions and authority, has escalated. Homosexual vernacular, style of dress, iconography, and ritual have been incorporated by a wider set of people, described by Crowley (1987:302) as *homosocial*. This feminist interpretation distinguishes between genital sexuality (homosexual) and an entire range of same-gender bonds (homosocial), and creates an identity hazard for some gays. They perceive the alternative, the androgynous, and even the bisexual as a threat to their style of sexual contacts, and as a fudging of rigid gay sub-cultural boundaries. The clues and cues of homosexual ritual are manifested by others whose sexual orientation is not necessarily homosexual. Gays nevertheless become seduced by external appeal substantiated by dress, language, and participation in a sub-culture, and a common experience narrated by gay people is frustration and indeed anger when sexual intimacy with 'alternative' heterosexuals is non-negotiable.

Alternative sub-cultures have blurred the gay sub-culture by taking co-ownership of aspects of it, such as the sharing of political ideologies which indicate a non-tolerance of oppression; the sharing of discotheques, coffee bars, and beaches; and the sharing of symbols, vernacular, and styles of dress (Hayes, 1981b).

These features, in conjunction with the limited availability of institutionalized gay ghettos, promote a sense of insecurity for many homosexual people. This is evidenced in the re-establishment of gay venues which provide 'safe' and exclusive outlets, such as private parties, and clubs and bars with gays-only patronage. The constant re-creation of an exclusive gay sub-culture in reaction to intrusion from non-gays has implications for identity crisis. It perpetuates the distancing of gays from others, thereby reinforcing their sense of separateness and estrangement from wider society.

9. Bisexuality has implications for homosexual identity

Bisexuality is not unfamiliar to most homosexuals. Sexual liaisons and romances with women constitute part of the identity experiences of many gays during their late teens and early adulthood (and indeed may continue throughout their lives). A clinical feature of these early bisexual experiences is a need to justify or prove 'heterosexuality' to others. However, once the homosexual identity continues to unfold, heterosexual intimacy ebbs in favour of consummating homosexual relationships.

Bisexuality is, however, a legitimate expression of sexuality for many people, and has emerged as an identity construct of its own. Paul (1984:56), in explaining sexual labels, not only draws attention to the vulnerability and confusion arising from a dual identity, but suggests that a person's experience of bisexuality is apt to be different based upon whether the bisexuality is *sequential* (varying from a solely homosexual to a solely heterosexual pattern, and back again), or *contemporaneous* (having male and female sexual partners during the same period).

In Chapter 1, bisexuality was accorded some status as a transitional, experimental phase (in identity terms), in which the individual attempts to negotiate his sexuality according to parameters set by society. Fantasy experiences of a homo-erotic nature are suppressed in favour of heterosexual experiences. Most homosexuals proceed to acquire an exclusively homosexual identity, but it must be stressed that bisexuality may constitute a prolonged lifestyle for many.

Bisexuality may be linked to the process of coming out where the individual gains homosexual status if and when the signals, cues, and appropriateness of owning a homosexual identity are strong enough. Marriage and/or the dating of women may be entered into to deflect from the homosexual feelings and behaviours, and for the sake of social approval. It is not uncommon for married men to have clandestine relationships with gay men.

In some of the research study commentaries on the nature of crisis in the coming out period, respondents attached great importance to the pain and fear of severing heterosexual relationships. Some had been engaged or even married. Bisexuality, and its association with coming out, is recognized as a period of emotional fraudulence, and carries the following protocols which make it especially crisis-prone:

▶ maintaining an appearance for the sake of social approval, specifically with regard to family, peer group, and work situations; and

▶ dealing concurrently with two sets of interactions; firstly having to respond to heterosexual cues, and secondly having to process, filter, and hide homosexual cues.

Because of the prevalence of bisexuality, notably during adolescence and early adulthood, the implications for AIDS are profound. Since bisexual behaviour is often part of the closet syndrome, the sexual intimacy of bisexuals is a special risk factor. In order for the person to actualize his homosexual and heterosexual desires, the need to maintain intimacy with both genders is a priority.

10. The state of the formal gay movement in South Africa is inextricably linked to identity issues

Chapter 6 described in some detail the state of the formal gay movement in South Africa, and earlier discussion has illustrated the power of such movements when strongly developed, as well as their contribution to gay identity. Moreover, there is extensive support in the literature for the notion that gay liberation is indeed connected with the liberation of oppressed minorities in general. Nevertheless, it can be suggested that homosexual identity cannot be fully actualized until gay rights are addressed. Unless South African gays are accorded the rights and privileges that endorse their status as citizens in good standing, identity issues will be problematic.

The Australian experience of homosexual liberation provides an illustration of how far South Africa lags behind some other countries in this regard.[1] South African homosexuals are disadvantaged in not having a powerful, national formal gay movement to promote gay rights. Some of the causes of this, which have implications for the collective gay crisis in South Africa, are that:

▶ There is no prominent gay leadership in South Africa at present.

▶ Gay organizations are primarily service-oriented, with their energies expended mainly on the AIDS issue.

▶ The formal gay movement, as represented by the Gay Association of South Africa (GASA), has dissipated as a result of political and social divisions. Politically, the majority avoid publicity and contentious action. Only the minority are activist, and through splinter non-racial groups such as GLOW and OLGA attempt to address gay issues in parallel with human rights. Negotiation with the African National Congress about their official stance towards homosexuality in a post-apartheid society is but one example of these attempts.

However, within the context of South African society, activist organizations could not until recently function without the imminent threat of government disapproval. Homosexuality, if linked by implication to the politics of liberation, would have had a subversive connotation attached to it. The majority of white gays are therefore inactive in the arena of liberation politics. Some further reasons for this are their preoccupation with their own coming out priorities, and racial attitudes (including prejudice) which have been part of their South African socialization.

It can be predicted with some certainty that until such time as anti-discriminatory measures are firmly entrenched in statute, gay identity, at both the individual and collective level, will remain in jeopardy.

Implications for intervention by helping professionals

The persistent theme that runs through this book is that crisis is associated with homosexual identity formation. It follows that a major task for members of the helping professions is to utilize effective ways of responding to these crises which cause such widespread and often long-lasting pain and confusion to homosexual people.

This section therefore considers a number of issues pertaining to crisis intervention as a response to homosexual crisis. The rationale for the use of crisis intervention methodology is discussed, and attention is then given to four particular aspects of working with homosexual clients in crisis: the use of fantasy in the helping process, transference and counter-transference, the need for helping professionals to have a set of appropriate attitudes towards homosexuality, and some specific intervention concerns with homosexual clients.

Why crisis intervention?

As an approach, crisis intervention is firmly entrenched in sound clinical practice, but its application to homosexual concerns has been minimal. A basic premise of crisis intervention is that professional helpers in community systems have a powerful influence on how individuals, groups, and families deal with and resolve crisis. The fact that crisis intervention arose out of a response to developmental, accidental, and life-transitional crises should place it well within the framework of professional practice in response to homosexual issues (Isaacs, 1979a, 1979b; Porter, 1966; Robertson, 1986).

The models of crisis intervention are based on the tenets of community and public health, a theme advanced by Caplan, the father of community psychiatry. A fundamental premise of crisis intervention is that crises are a natural and appropriate response to overly stressful predicaments that confront people. The mandate of intervention is to promote growth as a result of the crisis, and this consequently erases the label of ill-health from the experience. Lindemann's classic work of 1944 made clear that crises need the expression of appropriate emotions, in order to achieve the homeostasis required for further development and resolve. Subsequently, clinicians (such as Hollis and Woods, 1981) have pointed out that crises resurrect past events, often of a suppressed nature, and if dealt with timeously can prevent reactive responses in favour of proactive ones.

Viney states in regard to crisis that:

> The crisis concept avoids much of the pessimistic, devaluing, even invalidating approach we often make to patients, by viewing crises as part of normal development, by emphasising positive coping rather than negative defense manoeuvers, and by proposing crisis resolutions which allow for growth as well as regression (Viney, 1976:393).

Moreover, Baldwin suggests that the crisis intervention model, by virtue of its strong interdisciplinary character in both theory and practice, may be a unifying influence among various helping professionals. He goes on to remark that crisis intervention's ideas are congruent with the increasing emphasis on the interconnectedness of health and mental health care in treating *the whole person* (Baldwin, cited by Slaikeu, 1984:8).

Crisis intervention has had some exposure in South Africa. One of the writers started the first crisis clinic in South Africa in December 1971. The service closed down in June 1977 when the then Department of Social Welfare and Pensions (under whose aegis the clinic operated) questioned the so-called 'non-traditional' practices of the clinic. Subsequently, an attempt to establish a crisis clinic in Cape Town was abortive, but three new crisis units started in Johannesburg: the Waverly Crisis Clinic under the auspices of the Jewish community, the Randburg Crisis Clinic, and the Radio 702 Crisis Clinic. The latter clinic, under the overall auspices of the Witwatersrand Mental Health Society, is perhaps the most successful. It is directed by social workers and psychologists, and is supported by a team of appropriately trained volunteers.

The writers know of only two formal organizations in South Africa which deal specifically with homosexual crises from the perspective of the crisis intervention model. These are the Radio 702 Crisis Clinic, mentioned above, and the GASA 60-10 Counselling Centre in Cape Town. The absence of other formalized crisis services directed towards homosexuals merits further comment. Perhaps the major reason is that no national or regional South African body exists in the field of human sexuality. Unlike other areas, where national councils have been established to promote and co-ordinate activities in fields of social concern such as the physically disabled, cancer patients, tuberculosis sufferers, and the like, the plight of sexual minorities has not been addressed at policy and service-delivery levels. Homophobia, backed by legislation, has blocked all attempts in this regard. One wonders why, if homosexuality has been perceived as a 'condition' or 'deviancy', no attempts have been made to deal with this 'condition' in a way similar to that in which existing national councils and their affiliates have dealt with their specific concerns. While 'cure' is not advocated as a necessary objective in the establishment of such a council, it would have the value of a *legitimate* position in the health and welfare structure of the state and private initiative, and would be able to promote the appropriate use of helpful methodologies such as crisis intervention.

Gochros (1984) stresses the necessity of equipping helping professionals to meet the needs of the homosexually oriented, and together with Bohn (1984) and Messing et al. (1984) draws attention to the incidence of homophobic responses among mental health practitioners. De Crescenzo confirms this in her research findings, and alludes to the fact that, even though mental health workers are aware of contemporary dynamic thinking, they are likely to view homosexual people as immature, arrested in terms of sexual development, or neurotic by definition. She covers these attitudes under the broad definition of homophobia, which includes a fear of homosexuality (inculcated in early life), expressed antagonism, and oppressive legislation designed to eliminate homosexual behaviour (De Crescenzo, 1984:115–23).

Gonsiorek (1982) makes certain assumptions about clinical diagnosis, and believes that differential diagnoses (i.e. diagnoses of disease entities) are often misused in the framework of homosexual identity development. He points out that *sexual identity crises* are often confused with the coming out process, and regarded as a serious psychopathology. In fact, such a pathology does not exist. Rather, the individual is experiencing a partially (or at times completely) reality-based crisis as a result of forces such as severe interpersonal rejection, or the impending or actual loss of a job. This creates florid symptoms, since it may be perceived as frightening or ego-alien, thus contributing to subjective distress. Rusk (1971) refers to this as the crisis event, which is related to the inadequacy of the ego's adaptive and creative capacities to handle the stimulus of a change of input to the system. Anxiety signals increase, often overwhelmingly, and constitute a threat to ego-equilibrium and integrity. Hence, without obviously doing so, Gonsiorek highlights fundamental principles of crisis intervention. He specifically draws attention to the premise that the precipitant needs to be identified, for example by examining same-gender feelings and behaviour prior to the current crisis (Gonsiorek, 1982:18).

This embodies the principle of crisis intervention that addresses the genetic present as it relates to the genetic past. With crisis intervention, there is no single dynamic or pathogenic explanation. Indeed, Malyon believes that the initial diagnostic aim should be to delineate a tentative developmental profile that gives credence to the full range of formative variables (Malyon, 1982a:64), which will guide the helping professional to

▶ focus on the immediacy of the crisis;
▶ understand the development of the crisis;
▶ deal with the precipitant (which aids the assessment effort);
▶ facilitate re-experiencing the peak of tension in the crisis;
▶ isolate the critical (propitious) moment in order to deal with risk evaluation; and
▶ place the anxiety and fantasy expressions into a meaningful context for the person and the intervenor.

An overview of the theory and practice of crisis intervention and its specific application to homosexual crisis is set out in Chapter 2, while the use of this methodology with AIDS patients is discussed in Chapter 5. In the application of generic crisis intervention models to homosexual crises, certain aspects of the process require specific emphasis. The first of these is the use of fantasy in the helping engagement.

The use of fantasy in crisis intervention

Fantasies and images can provide helping professionals with direct access to a person's thoughts, wishes, feelings, and experiences. They can also be used to circumvent the defence mechanisms that prevent unconscious or repressed conflicts from surfacing. As Shorr (1983) notes, sexual conflicts deal with the most vulnerable, the most tender, the most shame-inducing, and the most guilty feelings, and consequently fantasies about sexual conflict are the most difficult to disclose to the self or others. According to Shorr, these fantasies include 'images of sexuality we have during intercourse or in place of it, as in masturbation, images related to strategies of interaction that anticipate sexual outcomes [and feelings of] dominance, rejection, jealousy, sin, joy, [and] being dirty' (Shorr, 1983:107).

Elphis (1987), Mann (1973), and Masters and Johnson (1979) specifically recommend goal directiveness (a function of crisis intervention) as the method of making conscious to the client the kinds and intensity of sexual fantasies that exist, in order to understand the central conflict. Coinciding with the basic premise of crisis intervention that the genetic past is captured in the genetic present, Salzberger-Wittenberg (1975:15) reminds us that fantasies are images or ideas that are based upon past patterns of relationships or experiences, and therefore they are transferred to the present.

Bearing in mind that crisis intervention deals with people who present as vulnerable and whose defence systems are usually at their lowest ebb, access to fantasies as a cue to psycho-sexual determinants of behaviour can be dealt with immediately (Isaacs, 1988). Shorr pays tribute to fantasy investigation in therapy, and believes that the imagery of the client facilitates the 'seeing' of what the client imagines, thus enhancing professional empathy. Shorr notes that intervention which deals with fantasy structures attempts to put the individual through his own imagery into a particular situation that may evoke a set of interactions, useful not only in revealing major problems but also in permitting him to relive painful or embarrassing experiences (Shorr, 1983:93).

Sexual fantasy is the core component of human sexuality. The reluctance of many helping professionals to explore the sexual fantasies of their clients has resulted in ignorance about the fantasy patterning of fully functional heterosexual, bisexual, and homosexual people. Thus cultural myths and stereotypes are perpetuated (Masters and Johnson, 1979). In order to substantiate their claim clinically, Masters and Johnson offer a guide to the content of fantasy material elicited from a sample of homosexuals, and list five major recurring themes evolved from 'free floating' descriptions. Although these categories are somewhat limited in their range of experiences, they offer helping professionals a guide to fantasy assessment. They are the following:

▶ imagery of sexual anatomy;

▶ forced sexual encounters (images of rape and seduction);

▶ cross-preference encounters (for example, switching roles from inserter to insertee, or passive to active);

▶ idyllic encounters with unknown men; and

▶ group sex experiences (Masters and Johnson, 1979:179).

Masters and Johnson warn helping professionals against becoming locked into a particular and static view of a person's fantasies, or an established interpretation of fantasy patterns. If and when fantasy patterns alter, so too must clinical interpretation and intervention. Therefore, although understanding of the client's inner world is central for any form of intervention (Rogers, 1988), it is also incumbent on the helping professional to recognize the particular frame of reference when dealing with fantasies. To this end the writers offer the classification in Chart 8.1, drawn from personal clinical experience, showing how fantasy systems can be utilized by helping professionals during the assessment and treatment phase of intervention.

The fantasy types, although presented discretely in Chart 8.1, are not mutually exclusive. Only for the convenience of diagnostic appraisal has each type been treated as if it were a separate category, for an individual may experience any combination of these fantasy types. For example, it is possible for a person who has reached the stage of *fantasy consolidation* to re-experience a *narcissistic period* as the result of some trauma (Rousso, 1985:14).

As a guide to the assessment and understanding of the growth of homosexual identity, it should be remembered that fantasies should not be seen as isolated from the total experience of the individual (Klein et al., 1985). Hence, fantasy exploration is but one feature of the assessment process (Slaikeu, 1984). Furthermore, some fantasies might indeed be both appropriate to, and necessary for, the person's stage of psycho-sexual development.

The types of fantasies reflected in Chart 8.1 may be experienced functionally or dysfunctionally. Examples of the latter include paraphile fantasies, such as necrophile images; sado-masochistic fantasies, incorporating violent images; paedophile fantasies where young children are the focal point of the fantasy content; and body

Chart 8.1 The range of fantasy types, their manifest content, and implications for intervention.

Fantasy type	Fantasy content	Denotation	Intervention strategy
1. Narcissistic	Focus on self-image, kinaesthetic arousal based on body mirror image; self is idealized	Self-involved, exclusive of others	Assist with identifying stage(s) of identity growth; locate sources of intimate patterns; gauge levels of self-esteem; identify patterns of trust
2. Fixed	Focus on externalized objects (both animate and inanimate) usually non-familiar and idealized; fantasy often repeats itself	Partial incorporation of external reality, but still idealized	Help to distinguish between open and closed boundaries, assist with identifying body or person types including the homosexual prototype
3. Experimental	Self with other objects (singular and collective); fixation of body parts; emphasis on homo-erotic images	Partial engagement in homo-eroticism	Locate sources of anxiety; elicit aspects of homosexual panic; identify levels of guilt and/or areas of comfort
4. Confused	Incorporation of both male and female objects, including transvestite and transsexual role play; fantasy has a voyeuristic quality; experiments with passive and active styles	Testing out of homo-erotic impulses in a dual context, i.e. bisexual phase	Identify bisexual confusion; examine levels of generalized confusion, panic, guilt and conflict; explore for possibilities of homosexual panic, and loss of masculine attributes
5. Dreams and images	Nocturnal dreams based on collective fantasy experiences, usually archetypal in nature, often linked to nocturnal emissions	Suppressed uncontrolled material which confirms *current* status of sexual identity	Determine levels of urgency and fixation on male imagery; suppressed or repressed material; identify panic responses related to the loss of control; gauge accurately ego-dystonic/systonic features
6. Learned	Fantasy images based on cult, body types, erotic zones, fashion, and sub-cultural iconography	Incorporation of sub-cultural image, indicates a partial or complete exposure to sub-culture and its various facets	Locate exposure/involvement with gay sub-culture; identify major sources of discomfort with self and others; deal with either *real* or imaginary experiences; highlight patterns of expressed or desired forms of intimacy; elicit power responses; gain access indirectly to sub-cultural world
7. Diverse	Multi-fantasy images; cerebral library of arousal; safer incorporation of images which are not necessarily idealized	Fantasies congruent with reality and identity, consequently divested of shame	Help identify later stages of homosexual growth; affirm identity baseline; deal with areas of control and safety
8. Consolidated	Fantasy usually reflects upon a significant and familiar person; ability to share fantasies with others	Fantasy complements sexual and emotional intimacy, high level of comfort with sexual identity	Locate sense of identity ownership; examine the absence or presence of fear, guilt and anxiety; anticipate future fears

distortion fantasies (self-distortion and distortion in others), which could lead to borderline psychotic episodes (Rosenbaum and Beebe, 1975).

The expression of these dysfunctional fantasies would require immediate attention. They have to be separated from the total crisis profile and dealt with as a priority issue. Fantasies are a clue to acting out behaviour, and the person might indirectly be warning the helping professional about aspects of his behaviour that have not as yet been revealed. Although the surfacing of dysfunctional fantasies may occur during a state of crisis, it must be remembered that the experience of crisis itself is not necessarily dysfunctional.

In accordance with crisis intervention, particularly when the anxiety level is maximal, the helping professional must mediate the catharsis by giving unconditional permission for fantasy expression. This assists with anticipating the outcome (a meta-fantasy experienced by client and helping professional alike), as well as dealing with the client's fears about 'contaminating' the helping professional with his ugly, strange, or different thoughts and behaviour.

Transference and counter-transference

The implications of the above have profound significance for both transference and counter-transference in the context of crisis intervention (Becker, 1988; Baptiste, 1987; Dunkel and Hatfield, 1986; Isaacs, 1979b; Malan, 1976, 1979; Mann, 1973; Orenstein, 1986; Rosenbaum and Beebe, 1975). Crisis intervention facilitates immediate access to transference and counter-transference during the assessment and intervention transactions. Furthermore, as a result of abandoning the traditional roles and taboos that exist between client and helper, the helping professional is in a unique position to deal immediately with the fantasy-anxiety-crisis triad. Ornstein, cited by Becker (1988:63), draws a distinction between the 'managed' or 'regulated' aspects of transference, as opposed to the dynamic understanding and interpretation thereof (i.e. actually dealing with the transference) which is the practice today among helping professionals.

Similarly, with respect to counter-transference feelings, the helping professional must at all costs distinguish his/her own set of moralities (which might include responses to legal proscriptions) and personal fantasies from the inner truth of the client.

It can be especially noted that while transference and counter-transference responses are an essential component of any therapeutic relationship, crisis intervention endorses the helping professional's *conscious use of self* as being therapeutically valid. Thus, transference responses are highlighted from the beginning of the therapeutic contact, with the primary goal of engaging both the client's and the therapist's feelings about each other. According to Isaacs (1979b:53–4), this point is significant within the intervention paradigm for many reasons. For example,

▶ it facilitates the contract system between client and therapist, and breaks down traditional role barriers, so that therapist and client 'labels' do not hinder the interactive component, and the 'sick' role of the client is minimized;

▶ it helps to deal with fantasies that the client may have about the therapist, by bringing them into the open; and

▶ it helps to strengthen and prepare the client for the termination process, which is characterized by issues of despondency, premature loss, abandonment, and rejection.

Since transference and counter-transference encompass unfinished business from the past as well as interactive components within the therapeutic relationship, both must be dealt with in the course of intervention. Crisis counselling within the context of gay concerns is symbolically linked to trust, acceptance, congruence, and sexual fantasies. The therapist becomes part of the 'rebirthing' process, particularly in identity and coming out matters, therefore symbolizing for the client an object of envy, desire, and idealization. These dynamics must be confronted in intervention.

Developing appropriate attitudes to homosexual issues

A further important variable is the attitude held by the individual helping professional towards homosexuality. Bearing in mind the discrepancy between clinical differential diagnosis and the professional hunches based on the development of crisis, Gochros and Schultz (1972) and Gochros (1984) invite helping professionals to establish a set of protocols when dealing with same-gender issues, which may involve change to their own attitudes. According to Gochros, in order to become effective in dealing with homosexual issues, the helping professional must prepare himself/herself in the following ways:

▶ locate sources of one's own personal discomfort;

▶ deal with personal fantasies that may or may not include homoerotic elements;

- differentiate between oppression and pathology;
- recognize the total personal identity of the homosexual;
- dissolve stereotypical responses towards gay people, and recognize their 'invisible' status;
- deal carefully with labelling, acknowledging particularly that while labelling simplifies the complexities of sexuality, it also obscures the wholeness of human beings;
- de-emphasize perceptions of separateness;
- gain perspective over the 'sickness' label; and
- develop and maintain links with gay people (Gochros, 1984:139–48).

The following case extract illustrates the discusssion in the three preceding sections, by demonstrating the therapeutic use of fantasy and transference, and the importance of the therapist having an appropriate set of attitudes towards homosexual issues:

A 20-year-old gay Technicon student approached one of the writers for help with a relationship problem. On presentation, he displayed feelings of worthlessness, a fear of loneliness, and anger towards the 'gay scene'. The client was known to the writer from another setting. This point was raised during the first session, in order to deal with issues of confidentiality and privacy, and to test out the level of embarrassment. Furthermore, it needed to be ascertained how both would feel should they meet in a situation outside of therapy, in order to deal openly with issues of comfort and avoidance.

As the therapy progressed, major concerns were identified. These included homophobia, parental rejection, fear of homosexual sex, uncontrolled camping, intimacy, and sexuality. It soon became evident that the therapeutic relationship was the client's first experience of a meaningful, safe, and potentially ongoing relationship. His sharing of intimate fantasies, sexual experiences, and his shame and guilt drew him closer to the therapist, but paradoxically also evoked strong feelings of anger. Upon exploration, it emerged that the client wished that he could have a lover like the therapist. In addition, the therapist's sense of 'therapeutic immunity' provoked images of distance and unavailability for the client. The client projected his vulnerability in negotiating his internal sexual imagery by making an ageist protest: he commented that he finds anyone older than himself [such as the therapist] unattractive in terms of personal relationships. Careful interpretation to the client of this transference response allowed him to express his disappointment (and relief) that the therapist would maintain his professional distance. Because it had surfaced and been identified, this transference response could then be linked to important dynamic areas, including those of dependency, trust, stereotyping, intimacy, betrayal, sexual power and roles, the meaning of relationships, and integrity and honesty in the expression of feelings.

With the transference responses, the usual pattern of dependency emerged, which the therapist linked to the termination effort by dealing with anxiety, loss (both actual and symbolic), abandonment, and unfinished business. In order not to 'disappoint' the therapist, the client embarked on a new social-sexual relationship, and proudly came to the last few sessions 'in love'. Further exploration revealed that the relationship was a reattachment (a rebound) in order to compensate for the impending loss of the therapeutic relationship, and to ensure that the therapist saw the client as whole and 'cured' (thereby sustaining the client's notion that gay relationships suggest wholeness).

Once this area was worked with in therapy, it emerged that the client still had difficulty in terminating. In order to help him with the ending, the therapist explained to him that maybe a 'door knob' issue would arise whereby new material would be brought into the last sessions as a reason for the therapy to continue. The client was assisted by the therapist using the technique of 'rehearsal for reality', whereby permission was given to him to feel safe to bring new material (if he so desired), thereby separating an actual concern/problem from the meaning of the relationship. In this particular case, the client came with HIV/AIDS issues, which were contracted into another three sessions, after which termination was ultimately achieved.

Other key issues in working with homosexual clients

Chart 8.2 (adapted from Slaikeu, 1984:19) is offered as an additional assessment tool for use by helping professionals when working with homosexual crises within a crisis intervention framework. Slaikeu proposes that the assessment, treatment, and evaluation functions of crisis intervention can be promoted by examining the individual in terms of five basic modalities (or sub-systems), in the context of his family, social group, community, and culture. This model de-emphasizes a focus on homosexuality as a 'condition', and directs attention to the client as a human being with areas of innate strength.

Chart 8.2 Basic personality profile

Modality/system	Variables/sub-system
Behavioural	Patterns of work, play, leisure, exercise, diet, sexual behaviour, sleeping habits, use of drugs, presence of suicidal, homicidal, or aggressive acts. Customary methods of coping with stress.
Affective	Feelings about any of the above; presence of feelings such as anxiety, joy, anger, guilt, depression. Appropriateness of affect to life circumstances. Are feelings suppressed or hidden?
Somatic	General physical functioning, health. Presence or absence of tics, headaches, stomach difficulties, general state of relaxation or tension, sensitivity of vision, touch, taste, sight, hearing.
Interpersonal	Nature of relationships with family, friends, neighbours, and co-workers; interpersonal strengths and difficulties. Number of friends, frequency of contact. Role taken with various intimates (passive, independent, leader, co-equal); conflict resolution styles (assertive, aggressive, withdrawn); basic interpersonal style (congenial, suspicious, manipulative, exploitive, submissive, dependent).
Cognitive	Current day and night dreams, mental pictures about past or future, self-image; life goals and reasons for their validity; religious beliefs; philosophy of life, presence of any of the following: overgeneralizing, catastrophizing, delusions, hallucinations, irrational self-talk, rationalizations, paranoid ideation, general (positive/negative) attitude towards life.

(Adapted from: Slaikeu, K. *Crisis Intervention: A handbook for Practice and Research*, Boston: Allyn and Bacon, 1984:119.)

The intrinsically unavoidable nature of crisis promotes the view that crisis in general is an essential part of human existence, and that all crises, if suitably dealt with by both victim and intervenor, are ultimately opportunities for some form of positive accommodation and growth (Isaacs and Miller, 1985:329).

Crises associated with homosexual concerns are no exception to this statement. Moreover, because homosexuality is not regarded as a pathology, no fundamental personality change is necessary when dealing with the crises pertaining to it. In *any* crisis, it is the person's *functioning* that is impaired, not the psyche or soma. Thus, the initial and continuing focus is one of prioritizing goals and marshalling the person's resources towards adapting to and meeting these goals, within the framework of his adaptive and functional capacities (Atkins et al., 1976:116).

From the diagnostic perspective, Atkins et al. (1976:123) warn all helping professionals who embark on a crisis intervention strategy with homosexual clients to make the distinction between whether the discussion of homosexuality is an expression of internal pain, or a request to change the self. The plea to helping professionals, therefore, is to examine the dichotomies presented by gays in crisis in respect of their egocentric and sociocentric profiles.

The foregoing discussion highlights vital features for helping professionals who work with homosexual clients. In combination with Slaikeu's 1984 multi-modal model, the views of Atkins et al. (1976) about intervention draw attention to key considerations:

▶ The term or diagnosis 'homosexual' is ineffective. It narrows the scope for examining the multi-dimensional experiences of the person. At the same time, it confines the helping professional and client to the restricting notion of homosexual behaviour only.

▶ Crises for homosexuals are in all fundamental ways similar to crises for heterosexuals. However, helping professionals should bear the following in mind when dealing with homosexual clients:

▷ some crises are unique to homosexuals; such experiences include a sense of crisis arising from legal and religious proscriptions, labelling, and homophobia; isolation as a result of rejection by the parent culture; the double bind of the homosexual sub-culture; and idiosyncratic sexual styles and intimacy needs;

▷ losses experienced by gays are compounded by the fear of expressing such losses, due to the non-legitimate status accorded to homosexuality by society;
▷ homosexual needs are met primarily through sexual affirmation, hence the issue of 'promiscuity' must be handled appropriately;
▷ relationships, whether transient or longer term, are sought to combat a sense of isolation and to confirm identity; and
▷ self-imposed isolation renders the person more vulnerable and may lead to serious depression.

Notes

1. The writers are indebted to Dr Gary Wotherspoon of the University of Sydney for the following account of gay liberation in Australia, which was given in the course of a personal interview:
Australian law in respect of human rights operates on a state, not a federal basis. Of the six states which comprise Australia, three have decriminalized homosexuality. The first was Southern Australia, which decriminalized homosexuality in 1972 following upon an inquiry into the much-publicized event of a gay academic being thrown into a river by three policemen. Eight years later, the state of Victoria, under a coalition conservative government committed to civil liberties, decriminalized adult homosexuality and reduced the age for consent between males to 16 years. In 1984, the state of New South Wales, which in its capital, Sydney, has one of the world's most densely populated homosexual communities, also decriminalized homosexuality, and fixed the legal age of consent between males at 18 years. The relative tardiness of New South Wales, compared to the other two states, is attributable to the former's strong Catholic tradition. The recent history of gay liberation in Australia dates back to World War 2. A flourishing 'Bohemian' and artistic set, in combination with members of the armed forces living in the cities, initiated friendship network groups which ultimately became the precursors of organized gay activism. Following this period, during the 'Cold War' of the 1950s and early 1960s, gays were regarded as subversive and hence as a threat to the state. However, the ideas of the 'new left', feminist protest movements, and person-centred politics became legitimized, leading to anti-discrimination laws. This period of law reform explicitly included

homosexual issues. Strong gay leadership gained momentum, and the gay movement, within the wider parameters of anti-discriminatory legislation, directed its efforts towards law reform, improving community education, protesting against the medical profession's interpretation of homosexuality as an illness, and bringing cases of discrimination before the courts.

Arising out of the liberation attempts, and in concert with the introduction of human rights legislation, the following rights are now recognized in the three Australian states of Southern Australia, Victoria, and New South Wales:

▶ No one can be denied equal access to services, or equal service delivery.

▶ Bond and housing loans are non-discriminatory.

▶ No discrimination in the place of work is allowed.

▶ The federal government recognizes the legitimacy of homosexual relationships.

▶ Superannuation benefits are passed on to lovers.

▶ Benefits allowed to married couples are allowed equally to gay couples (for example, the national airline offers the same fare reduction benefits to gay couples as to heterosexual married couples).

Appendix A

A case study of a person with AIDS: Robert

(*Note*: For the purposes of client protection and privacy, aspects of 'Robert's' identifying data have been changed. This case study has been taken from the clinical records of one of the writers, and is used to illustrate the therapeutic process in combination with the crisis intervention strategies described in Chapter 5.)

Background

Robert, a white male aged 39 years, was referred to one of the writers by the consultant at a psychiatric hospital. Robert, at the time of referral, was hospitalized for serious depression, and for investigations into his neurological status.

The first contact with Robert at the hospital was undertaken conjointly with the psychiatric registrar who had been treating him. The major purpose of this session was to introduce Robert to one of the writers in his clinical role, and to provide a synopsis of the patient's current medical status in the presence of the patient himself.

Robert presented as thin, fragile, extremely depressed, and tearful. During the session, he constantly referred to his physical discomfort and sought remedial answers from the psychiatric registrar. It was clear that the physical ramifications of AIDS were the patient's primary disturbance at the time, and that therapy would initially have to deal with the anxiety and realities pertaining to the physical manifestation of AIDS-related illness.

Some of the salient features of Robert's immediate history were the following. He was admitted to the hospital on the advice of a Cape Town physician. A primary reason for the admission was 'severe depression, melancholia, and possible neurological damage'. A previous episode of pneumonia served to confirm an AIDS diagnosis. Furthermore, a mild epileptic attack, with the patient complaining of short-term memory loss, precipitated the hospitalization.

Robert had been diagnosed as HIV-positive the year before, when he presented with recurring symptoms of swollen glands, fatigue, weight loss, night sweats, and general malaise. His live-in lover, Peter, a man aged 22 years, had also been diagnosed as HIV-positive, and had died prior to Robert's admission. They had been together for nearly four years. Robert previously worked as a travel representative, holding down a senior position, living abroad for two years prior to meeting Peter.

His family, comprising parents and three siblings, lived in another part of South Africa, and both parents were ailing. With the exception of one brother who lived in the Cape, there was minimal contact between Robert and his family. He had been unemployed for two years, and was currently in receipt of a state disability grant.

Medical status at first contact

A profile of Robert's medical history (as pertaining to AIDS and AIDS-related illness) included the following:

▶ No clinical evidence of myelopathy or brain damage revealed by CAT scans and EEG investigations.

▶ Severe night sweats, swollen and painful lymph glands, and persistent diarrhoea, with concomitant weight loss.

▶ Continual bleeding and discharge from the rectum, which had not responded to any form of medical intervention.

▶ The emergence and persistence of oral thrush (candidiasis).

▶ Manifestation of sores, and bruise-like mulberry eruptions on his legs (which had recently disappeared, thus ruling out Kaposi's Sarcoma).

▶ Feelings of paraesthesia in the lower extremities (hands and feet).

▶ Difficulty with sleep, and interrupted sleep patterns.

▶ A definite diagnosis of AIDS-related complex had been given, based *inter alia* on evidence of pneumocystis carinii as confirmed by X-rays and clinical laboratory findings.

Robert's current psychological state included reactive depression, lack of hope, guilt, and the beginnings of despondency. His major fear was his inability to regain control over his body and possible loss of mental functioning.

He expressed relief and a keen desire to engage in a regular therapeutic relationship with his social worker (i.e. one of the writers). A therapeutic contract was negotiated, in which Robert would be seen bi-weekly at the hospital, and where agreement was made that the therapeutic team would be informed of certain issues arising from the therapy in order to facilitate his comfort in the hospital. In this respect, it must be noted that a fear expressed by patient and staff alike was other patients' responses to Robert should they discover the nature of his illness, as well as some form of barrier nursing and privacy as a result of his infectious state.

Psycho-social overview

The salient features to emerge from Robert's history, in terms of the homosexual aspect of his personality, were as follows:

Robert, having been raised in a strict Afrikaans patriarchal tradition, with strong religious influences, initially responded with fear and trepidation to his homosexual identity, which became apparent to him at the age of 13. After matriculating, he separated himself geographically from his family, and began to negotiate his sexuality within the homosexual sub-culture. When his first serious relationship ended, he confronted his family with his alternative lifestyle, and eased into a comfortable relationship with them once they had reconciled their own fears about homosexuality. With tacit family support, he had no difficulty in submerging himself in his lifestyle. Although he had experienced many different homosexual liaisons, his objective was to settle into a long-term relationship. He had three love relationships, all of which were, according to him, meaningful. The termination of each relationship was painful for him, but he rebounded and reattached himself to significant others almost immediately after each relationship ended. During his sojourn overseas (in the USA), he experienced intimacy with several people, and imbibed alcohol, dagga, and amyl nitrite ('poppers'). He expressed no guilt or discomfort about his choice of lifestyle, and believed that the gay scene nurtured him and offered him the support and excitement that he needed. Upon his return to Cape Town, he met Peter, and they soon became lovers. Their relationship, although intense and satisfactory, was not without conflict. Arguments persisted, with Peter on several occasions leaving home for two or three days at a time. Robert believed it to be partly due to the disparity in their ages, and the need for Peter to explore his belated sense of adolescence, an experience which had been suppressed due to 'closet' behaviour. When they were both diagnosed simultaneously as being HIV-positive, their relationship became stronger and was tempered with a compassion that neither had experienced before. When Peter died, Robert happened to have been away from Cape Town attending a family function. He arrived home to find his lover dead. This led to depressive behaviour and coincided with his deteriorating physical condition, which in turn led to his hospitalization and contact with one of the writers.

The clinical résumé

The first sessions were concerned with Robert's intense preoccupation with Peter's death. He reminisced about their relationship, and lived in expectation that Peter would return to life. Part of this feature was compounded by the fact that he had not been present at Peter's death, and unfinished business remained. Furthermore, this condition was reflected in Robert's obsessive visits to Peter's grave, where he 'spoke' to Peter, and asked for forgiveness. A series of photograph albums, a portfolio of their relationship including poems and letters written between the two lovers, were brought into the therapeutic sessions. This enabled Robert to share visibly with his therapist the pleasures and anguish of the relationship, thereby facilitating the mourning process. Peter had been idealized by Robert, so that Robert felt not only guilt, but also anger towards Peter's death and AIDS. During these sessions, Robert's depression and sense of discomfort was palpable, and surfaced easily. The chief dilemma was to partialize and control the flood of variable emotions confronting both the client and the therapist. Each session had to deal with

- ▶ the extent and relevance of the AIDS symptoms, including Robert's obsession with newspaper cuttings about AIDS, which he had meticulously filed;
- ▶ the physical discomfort which Robert was experiencing, and his anger towards the medical profession for being unable to treat the symptoms successfully;
- ▶ his fear of being rejected by staff and patients;
- ▶ his feelings of failing his family, and not being able to support his ill parents;
- ▶ his feelings towards his current flatmate who, Robert felt, did not understand his predicament; and
- ▶ anticipation of discharge from the hospital, and how he would cope financially with no more than a disability grant on which to live.

During these sessions, the compounding features of Robert's situation became increasingly apparent. Not only was there the burgeoning issue of AIDS and its consequences with which to contend, but the areas of sexuality, intimacy, and unresolved business had to be dealt with too. Thematic contracting (dealing with priorities) was almost impossible due to Robert's variable physical responses from day to day. As a result, anger was directed towards the therapist for not being able to 'solve the problem'. Direct feelings such as 'Why can't you tell me what to do', and 'You don't seem to understand', began to intrude upon the sessions.

In order to defuse the pace of the anguished present, the therapist encouraged Robert to retrace his history, so that features of the genetic past could be related to the immediate present. Through reminiscing, strengths of Robert's past were introduced into the sessions. This reconstruction of the past was used to infiltrate Robert's apparent reluctance to seek out cause from effect. He absorbed the entire blame for Peter's death and disallowed any appropriate feelings about himself. This eventually led to the discovery that neither he nor Peter could be definitively accused of causing one another's illness. Prior to their meeting, both had experienced multiple sexual partners, and there had been no guarantee of fidelity during their relationship. These factors emerged only towards the latter part of the sessions. Furthermore, both were anal recipients as well as anal inserters in relationships. The risk factor was highlighted to him, without resorting to blame but by referring to circumstance. The idealization of Peter began to recede, and the realities of the relationship began surfacing. Permission for Robert to mourn (specifically to cry) was frequently spelled out. As the anger, grief, and disappointment emerged, so the depression started to lift. Features of so-called memory lapse were obvious. It was put to him that, rather than the manifestation of clinical memory loss, suppression of painful issues was apparent, and that therapy provided the safety for him to express these fears.

The ensuing sessions took advantage of his lessened depression, and began to explore his AIDS profile. Fear of never again being able to capture a sense of sexual intimacy with others was the focal point. Anger towards the gay scene was expressed, and manifested itself as a reaction formation syndrome. All that had given him pleasure was now perceived as a potential container for disease. This defence was readily dealt with by the therapist. Thereby Robert was allowed to discuss his gay behaviour in perspective, and to pave the way for alternative behaviour within the context of his present disability. Intimacy would have to be achieved by renegotiating sexuality into other areas of personal action. He began to relate to the staff of the hospital and, following the advice of the therapist, sought to involve himself with a member of the 60-10 group who had embarked on an AIDS support programme. When Robert felt embarrassed or ashamed, particularly when socializing with friends, he was encouraged to appraise his feelings and to share them with his friends, thus paving the way for alternative but equally important forms of intimacy.

The last sessions with Robert (prior to a temporary six-month termination when the therapist would be away from Cape Town) dealt with his discharge from the hospital, his feelings about being 'abandoned' by the therapist, and reformulating his priorities. Plans for him to set up a stall at a local daily market venue were debated, medical continuity was discussed, and a locum tenens therapist was appointed.

Upon return after six months, contact was made with Robert. Severe deterioration had occurred, and both his physical condition and mental stability had declined. Features such as his rectal bleeding, night sweats, painful lymph glands, weight loss, and severe depression continued. His affect was flat and blunted, and sessions were marked by heavy crying and fatigue. The major factor that emerged was his preoccupation with death. Although he could not verbalize it *per se*, the symbols of *Thanatos* were apparent.

One example is relevant. Robert described a recent incident of fetching a friend from the airport. He narrated in a perplexed manner how he had got lost and landed up in Crossroads, a squatter camp, late at night. His verbalization of his fear and panic was acute, but at the same time a sense of amazement was evident at the ostensible peace and calm within the boundaries of the camp. He managed to leave the camp, and drove to the airport to await the arrival of his friend. The therapist acknowledged the difficulty which he experienced in speaking about this issue, and ventured to interpret that, alongside his other statements of 'getting papers in order, and throwing things away', he was unconsciously disappointed with his experience of Crossroads. He was hoping to be confronted by an angry mob who would set his car aflame, and so end it all for him. Robert responded with relief, confirming that he was now plagued with ideas of self-destruction but, because of glimmers of hope, he could still survive. It was suggested to him that the very act of surviving led to 'survivor guilt', for he compared himself with his late lover and two friends of his who had recently died. In effect, the crisis of choice was available to him. Therapy would be geared to dealing with the nature and consequences of choice. Choice was also interpreted to him as a valuable human phenomenon, and that, as long as he retained his capacity to choose, some levels of his ego integrity would remain intact.

Suicide ideation needed to be expressed, for this is a natural phenomenon of impending deterioration. In the context of death preparation, or anticipated long-term suffering, factors such as the quality, as opposed to quantity of life were discussed. Robert was able to verbalize that it was not the fear of death that troubled him, particularly as he felt a sense of religious fortitude, but rather a sense of lack of abatement of his physical and mental anguish. External stress factors emerged, such as lack of money and decreasing support systems, coupled with envy directed towards healthy people (specifically those in relationships), which compounded his feelings of generalized despondency. A new concept of his self-image was discussed, particularly as at this stage negotiating any form of sexuality seemed out of the question.

Of concern to both Robert and the therapist were his feelings of loss of control. More directly, his physical handicap was intensifying alongside his emotional lability. One point of therapeutic significance highlighted during the sessions was his ability to be totally honest with the therapist, and so to negotiate a new sense of intimacy. Although this placed a heavy burden on therapeutic transference and counter-transference, the outcome of this intimacy could be used by Robert during his interactions with others. A strong limiting factor which seemed to reduce his sense of external drive was the fact that he daily grew more despondent, physically and psychologically.

Concluding observations and recommendation

The therapeutic objective at this stage, although contrary to conventional mental health principles, was once again to remove Robert from his immediate community and readmit him to the hospital. This was fraught with contra-indications, for symbolically it could have suggested an inability in Robert to cope as an independent human being, thus promoting his sense of deterioration. On the other hand, he was reaching a point of such fatigue that he needed the caring and companionship offered by the hospital. Of note was the fact that his previous experience of hospitalization represented a period of tranquillity for him, and he was able to recall the inner peace that he achieved there. Robert was encouraged to decide himself on his need to be hospitalized, thus reinforcing any sense of integrity that he retained.

Appendix B

Survey questionnaire

Confidential questionnaire

Age: _____ years

Area of Residence (e.g. Sea Point, Stellenbosch, Bellville):_____

Present Occupation (e.g. businessman, teacher, unemployed, electrician, surgeon, university student, etc.):_____

Question 1

At approximately what age were you first aware of feelings of physical attraction towards persons of the same sex?

_____ years

Comment (if any): _____
_____
_____
_____
_____

Question 2

At approximately what age did you 'come out'? (Please see sheet of definitions)

_____ years

Comment (if any): _____
_____
_____
_____
_____

Question 3

Did you experience a crisis in respect of your 'coming out'? (Please see sheet of definitions)

_____ Yes
_____ No
_____ Uncertain

Comment (if any): _____
_____
_____
_____

Question 4

If you did experience a crisis, was it an unpleasant or disturbing experience for you?

_____ Yes
_____ No
_____ Uncertain

Comment (if any): _____
_____
_____
_____

Question 5

How did you 'come out'? (Tick off as many responses as are applicable in your case)

_____ It just happened
_____ It was forced on you
_____ You sought out other gays (in bars, clubs, etc.)
_____ You read about homosexuality
_____ People talked to you about it
_____ Through therapy

Some other way (Please specify: _____
_____)

Comment (if any): _____
_____
_____
_____
_____

Question 6

Apart from any crisis you may have experienced in respect of your 'coming out', are there any other aspects in which being gay has caused a crisis in your life?

_____ Yes (please specify: _____
_____)
_____ No
_____ Uncertain

Comment (if any): _____
_____
_____
_____
_____

Question 7

Do you think that other gay people, whom you know or have met, have experienced crises in 'coming out'?

_____ Yes
_____ No
_____ Uncertain

Comment (if any): _____
_____
_____
_____
_____

Question 8

Do you think that gay people in general think of themselves as significantly different from 'the man in the street'?

_____ Yes (please specify: _____
_____)
_____ No
_____ Uncertain

Comment (if any): _____
_____
_____
_____

Question 9

Would you like to see a centre established that would cater specifically for persons who have concerns with their homosexuality?

_____ Yes
_____ No
_____ Uncertain

Comment (if any): _____
_____
_____
_____

Question 10

Tick those statements below about the 60-10 group that are true for you.

_____ It could have helped me in 'coming out'
_____ It alleviates loneliness
_____ It facilitates meeting people
_____ It bores me
_____ I dislike it because it caters for gays only
_____ I think its members are too snobbish

_____ It could provide information relevant for gay people
_____ Other (please specify: _____
_____)

Comment (if any): _____
_____
_____
_____

Question 11

Would you like to help gay people in distress?
_____ Yes
_____ No
_____ Uncertain

Comment (if any): _____
_____
_____
_____

Question 12

If so, how would you like to offer your help? (Tick all that apply)
_____ on your own
_____ via the 60-10 group
_____ through some other formal organization
_____ through some other channel (please specify: _____
_____)

Comment (if any): _____
_____
_____
_____

Question 13

Do you feel that the request to fill in this questionnaire is an infringement on your privacy?

_____ Yes
_____ No
_____ Uncertain

Comment (if any): _____
_____
_____
_____
_____

Thank you for your co-operation and assistance

Sheet of definitions

By *crisis* is meant a serious situation or event in your life accompanied by, for instance

▶ a feeling of panic;
▶ intense feelings of fear or anxiety;
▶ a threat to your well-being;
▶ a sense of loss;
▶ the feeling that you are no longer able to cope properly with your everyday tasks.

By *coming out* is meant your recognition that you are homosexual rather than heterosexual. This may happen via an event or situation, or an accumulation of feelings and/or experiences, whereby you admitted to yourself (and possibly to others, too) that you are gay.

Appendix C

Glossary of gay vernacular in use in South Africa

AC/DC	bisexual
active	a person who is sexually active; usually with the underlying assumption that his sexual preference is anal penetration/intercourse
Ada	the buttocks area, usually in relation to base relief in tight pants
affair	two people in a homosexual relationship
Agatha	a person who gossips, usually maliciously
aggie	homosexual gossip
aggro	aggression; an aggressive person
AIDA	a person with AIDS
belenia	a physically attractive man
Bella	to be beaten up; a person who beats up homosexuals
Betty	buttocks; size of the buttocks
Beulah	an aesthetically beautiful man
bit	general description of a homosexual
bitch	person with a sharp tongue; person who causes strife within a clique
bliss	a person or situation that arouses physical, erotic, or social excitement
blow job	fellatio (sucking of penis)
B.M.	a heterosexual person (abbreviation of 'baby-maker')
butch	masculine; masculine qualities in a man, as opposed to feminine traits
camp	(*adjective*) a person who dresses or behaves in a manner which advertises his homosexuality; also refers to extravagant or 'kitsch' decor, clothes, jewellery, etc.
camp	(*verb*) courting, or hustling behaviour to attract attention, specifically to attract a person with the intention of becoming sexually involved
Celia	offering a cigarette to a passer-by with the intention of 'camping' him
chicken	young boys who prostitute themselves
chubby chaser	a person physically attracted to fat or extremely large men

Clora	a 'coloured' homosexual
closet	a person who has, or who is thought to have homosexual tendencies, and who has not acknowledged them (hence the phrase *in the closet*)
cock	penis
come	(also *cum*) ejaculation of semen
coming out	acknowledging homosexual identity, and commencing participation in the homosexual world
Connie	the moment of orgasm/ejaculation
Cora	a person who is common
cottage	a lavatory frequented by homosexuals in order to indulge in sexual acts
Debra	depressed, usually used in relation to depression over a relationship breakup
Delia	an excessively dramatic person (also known as a drama queen)
Dora	alcohol; hence the phrase *he is dora'd* for 'he is drunk'
Doris	an alcoholic drink
drag	female attire worn by a male, whether in fun or because of transvestism
dyke	a lesbian
Ethyl	elderly
faggot	a derogatory term in South Africa (but not in the USA) for a homosexual man
fag hag	a heterosexual female who moves almost exclusively in male homosexual circles
family a	homosexual
femme	a person who is sexually and emotionally passive and 'feminine' in a homosexual relationship
Fiona	wanting to have intercourse with a man; to have intercourse
freak	a person in the homosexual context who is radical and non-conformist, particularly in dress and attitudes towards society
F.S.Q.	a physically beautiful person, with 'film star qualities'
Gail	to talk; to have a chat
gay	generic and widely accepted term to describe male and female homosexuals
high camp	(see also the adjective *camp*) obvious homosexual behaviour i.e. dress, body movements, the content of a movie, home decor, etc.

Hilda	a physically plain and unattractive person
Iris	an Indian homosexual
Laura	a lover
lettie	a lesbian
Lily	the law; the police
lunch	the male genitalia
Marie	an eccentric or 'mad' person
moffie	a male homosexual (originally a pejorative term)
Monica	a woman having a menstrual period; alternatively a teasing term for a 'femme' person
MSM	men who have sex with men
Natalie	a black homosexual
N.O.C.D.	'not our class, dear'; referring to people of alleged lower class, status, or intellectual ability
Nora	an unintelligent person
number	a person to whom one is either attracted, or sexually involved with
Olga	to be organized
Olive	a very beautiful man
passive	a person who prefers to have anal intercourse practised on him
Penelope	to urinate
picnic basket	shape and size of penis and testicles outlined within trousers, underwear, or bathing suit
piece	a homosexual person
Polly-paranoia	a person with a poor self-image and lacking self confidence
pomp	to have anal sexual intercourse
pouff	a derogatory term for a homosexual, usually used by non-homosexuals
Priscilla	a police officer; police officers; 'the law'
p.y.t.	pretty young thing
queen	a homosexual who has absorbed female characteristics into his everyday behaviour
queer	as a *noun*, a homosexual person; as an *adjective*, homosexual
Reeva	revolting
rent	a male sex worker; or a homosexual who charges for his services
Rita number	a homosexual who charges for sex, usually with older men
Sally	to perform fellatio, to *suck off* another

scene	the homosexual sub-culture; hence *being on the scene* when a person is known among his peers as a practising homosexual
S. & M.	sado-masochistic; people who belong to an exclusive sub-culture involving leather fetish and painful sexual actions
Stella	a dishonest person; one who steals
Suzy	to tease a person playfully, to *send him up*
T.B.H.	'to be had'; an ostensibly heterosexual person with the potential to be homosexually seduced
T.D.F.	'to die for'; a physically beautiful man
T.D.S.	tedious; boring
Tilly toss off	to masturbate, alone or mutually
T.O.Q.	'tatty old queen'; an older, lecherous gay man
trade	a casual, transitory sexual experience
Ursula	a heterosexual person sympathetic to homosexuals
wada	to point out or stare at an attractive person
wank	masturbate
Wendy	a white homosexual
willy	penis

Bibliography

ACKROYD, Peter (1979) *Dressing Up: Transvestism and Drag. The History of an Obsession*. Great Britain: Thames and Hudson.
ADLER, Michael W (1987) 'Development of the Epidemic — ABC of AIDS', *British Medical Journal*, 294, 1 083–5.
ADREY, Robert (1970) *The Social Contract. A Personal Inquiry into the Evolutionary Sources of Order and Disorder*. London: Collins.
AGUILERA, Donna C and **MESSICK, Janice M** (1978) *Crisis Intervention: Theory and Methodology*. Saint Louis: C V Mosby.
ALTMAN, Dennis (1981) *Coming Out in the Seventies*. Boston: Alyson Publications.
ALTMAN, Dennis (1986) *AIDS and the New Puritanism*. London and Sydney: Pluto Press.
AMERICAN PSYCHIATRIC ASSOCIATION (1980) *Diagnostic and Statistical Manual of Mental Disorders*, Third Edition. Washington, DC: APA.
ANDERSON, Craig L (1982) 'Males as Sexual Assault Victims: Multiple Levels of Trauma'. In: Gonsiorek, John C (Ed.) *Homosexuality and Psychotherapy — A Practitioner's Handbook of Affirmative Models*. New York: Haworth Press, 145–63.
ARGUS, The (1984) 11 August.
ATKINS, Merilee; FISCHER, Mary; PRATER, Gwen; WINGET, Carolyn and **ZALESKI, Joanne** (1976) 'Brief Treatment of Homosexual Patients', *Comprehensive Psychiatry*, 17 (1), 115–25.
BABUSCIO, Jack (1976) *We Speak for Ourselves — Experiences in Homosexual Counselling*. Philadelphia: Fortress Press.
BALDWIN, B A (1979) 'Crisis Intervention: An Overview of Theory and Practice', *The Counselling Psychologist*, 8, 43–52.
BAPTISTE, David A (1987) 'Psychotherapy with Gay/Lesbian Couples and Their Children in "Stepfamilies": A Challenge for Marriage and Family Therapists', *Journal of Homosexuality*, 14 (112), 223–38.
BARKER, Philip (1983) *Basic Child Psychiatry*, Fourth Edition. London: Granada.
BARTOLUCCI, G and **DRAYER, C S** (1973) 'An Overview of Crisis Intervention in the Emergency Rooms of General Hospitals', *American Journal of Psychiatry*, 130, 953–60.
BAUM, Rudy M (1987) 'The Molecular Biology: AIDS', *Chemical and Engineering News*, 65 (47), 14–26.
BAUMEISTER, Roy F; SHAPIRO, Jeremy P and **TICE, Dianne M** (1985) 'Two Kinds of Identity Crisis', *Journal of Personality*, 53 (3), 407–20.
BAYER, Ronald (1987) *Homosexuality and American Psychiatry. The Politics of Diagnosis*. (With a new Afterword on AIDS and homosexuality.) Princeton, New Jersey: Princeton University Press.

BECKER, Lily (1988) 'Brief Dynamic Psychotherapy. An Exploration of Attitudes and Practice among a Group of Local Clinicians — Some Implications for Training'. Unpublished Master's dissertation (clinical social work). University of Cape Town.

BELL, A (1965) 'Research in Homosexuality: Back to the Drawing Board', *Archives of Sexual Behaviour*, 4 (4), 421–31.

BELL, Alan P (1975) 'The Homosexual as Patient'. In: Green, Richard (Ed.) *Human Sexuality. A Health Practitioner's Text.* Baltimore: Williams and Wilkins, 55–72.

BELL, Arthur (1984) 'The Sixties'. In: Denneny, Michael; Ortleb, Charles and Steele, Thomas (Eds.) *The View from Christopher Street.* London: Hogarth Press, 27–9.

BELL, Alan and **WEINBERG, Martin S** (1978) *Homosexualities. A Study of Diversity among Men and Women.* New York: Simon and Schuster.

BELLAK, L and **SMALL, L** (1965) *Emergency Psychotherapy and Brief Psychotherapy.* New York: Grune & Stratton.

BELLAK, Leopold and **FAITHORN, Peri** (1981) *Crises and Special Problems in Psychoanalysis and Psychotherapy.* New York: Brunner/Mazel.

BEM, S L (1974) 'The measurement of Psychological Androgyny', *Journal of Consulting and Clinical Psychology*, 42, 155–62.

BERGER, P L and **LUCKMAN, T** (1966) *The Social Construction of Reality.* New York: Doubleday.

BERGER, Raymond M (1982a) *Gay and Grey: The Older Homosexual Man.* Urbana: University of Illinois Press.

BERGER, Raymond M (1982b) 'The Unseen Minority: Older Gays and Lesbians', *Social Work. Journal of the National Association of Social Workers*, 27 (3), 236–42.

BERGER, Raymond M (1983) 'What is a Homosexual? A Definitional Model', *Social Work. Journal of the National Association of Social Workers*, 28 (2), 132–5.

BERZON, Betty (1979) 'Developing a Positive Gay Identity'. In: Berzon, B and Leighton, R (Eds.) *Positively Gay.* Millbrae: Celestial Arts, 1–14.

BERZON, Betty and **LEIGHTON, Robert** (Eds.) (1979) *Positively Gay.* Millbrae, California: Celestial Arts.

BEVERLEY, Peter and **SATTENTAU, Quentin** (1987) 'Immunology of AIDS/ABC of AIDS', *British Medical Journal*, 294, 1 536–8.

BIEBER, Irving (1962) *Homosexuality: A Psychoanalytic Study.* New York: Basic Books.

BIEBER, Irving (1972) 'Homosexual Dynamics in Psychiatric Crisis', *American Journal of Psychiatry*, 128 (10), 88–92.

BIGGAR, J (1986) 'The AIDS Problem in Africa', *The Lancet*, i, 79–82.

BIGSBY, C W E (Ed.) (1976) *Approaches to Popular Culture.* London: Edward Arnold.

BILLER, Henry B (1971) *Father, Child and Sex Role. Paternal Developments of Personality Development.* Massachusetts: Heath Lexington Books.

BLACHFORD, Gregg (1981) 'Male Dominance and the Gay World'. In: Plummer, Kenneth (Ed.) *The Making of the Modern Homosexual*. London: Hutchinson, 184–210.

BOHN, Ted R (1984) 'Homophobic Violence: Implications for Social Work Practice', *Journal of Social Work and Human Sexuality*, 2 (2/3), 91–110.

BOONZAIER, Emile and **SHARP, John** (Eds.) (1988) *South African Keywords: The Uses and Abuses of Political Concepts*. Cape Town and Johannesburg: David Philip.

BORLAND, Mary C; TASKER, Mary; EVANS, Patricia M and **KERESZTES, Judith S** (1987) 'Helping Children with AIDS: The Role of the Child Welfare Worker', *Public Welfare*, Winter Issue, 2 329.

BOSWELL, John (1980) *Christianity, Social Tolerance and Homosexuality. Gay People in Western Europe from the Beginning of the Christian Era to the Fourteenth Century*. Chicago: University of Chicago Press.

BOWLBY, John (1970) Attachment and Loss, 1 (Attachment). London: Hogarth Press.

BOZETT, Frederick W (1981) 'Gay Fathers: Evolution of the Gay-Father Identity', *American Journal of Orthopsychiatry*, 51 (3), 552–9.

BRACHT, Neil F (1978) *Social Work in Health Care. A Guide to Professional Practice*. New York: The Haworth Press.

BRAMMER, Lawrence M (1985) *The Helping Relationship: Process and Skills*, Third Edition. New Jersey: Prentice-Hall.

BRONSKI, Michael (1984) *Culture Clash — The Making of Gay Sensibility*. Boston: South End Press.

BROWNING, Christine (1987) 'Therapeutic Issues and Intervention Strategies with Young Lesbian Adults: A Developmental Approach', *Journal of Homosexuality*, 14 (1), 45–52.

BUCKINGHAM, Stephan L and **VAN GORP, Wilfred G** (1988) 'Essential Knowledge about AIDS Dementia', *Social Work. Journal of the National Association of Social Workers*, 33 (2), 112–14.

BUIS, Robert P D (1979) 'The Relationship between the Dogmatic Teachings and Attitudes towards Race Relations in two South African Religious Denominations'. In: Hare, A Paul; Wiendieck, Gerd and Van Broembsen, Max (Eds.) *South Africa: Sociological Analyses*. Cape Town: Oxford University Press, 86–112.

BULLOUGH, Vern L (1979) *Homosexuality — a History from Ancient Greece to Gay Liberation*. New York: New American Library.

BUNDY, Colin (1987) 'Street Sociology and Pavement Politics: Aspects of Youth and Student Resistance in Cape Town, 1985', *Journal of Southern African Studies*, 13 (3), 302–30.

CADY, Joe (1980) 'Coming Out'. In: Picano, Felico (Ed.) *A True Likeness: Lesbians and Gay Writing Today*. New York: The Sea Horse Press, 188–9.

CALDWELL, J M (1967) 'Military Psychiatry'. In: Freedman, A N; Kaplan, H I and Kaplan, H S (Eds.) *Comprehensive Textbook of Psychiatry*. Baltimore: William & Wilkens, 1 183–7.

CALHOUN, Lawrence G; SELBY, James W and **KING, Elizabeth** (1976) *Dealing with Crisis: A Guide to Critical Life Problems.* Englewood Cliffs, New Jersey: Prentice-Hall.
CAMERON, Edward (1985) *Untitled Talk delivered to GASA Convention.* Johannesburg, 31 May 1985.
CAPE TIMES, The (1978) 27 May.
CAPE TIMES, The (1980) 28 July.
CAPE TIMES, The (1986) 22 April.
CAPITAL GAY (1987) 18 September.
CAPLAN, G (1961) *An Approach to Community Mental Health.* New York: Grune and Stratton.
CAPLAN, G (1964) *Principles of Preventative Psychiatry.* New York: Basic Book.
CARKUFF, R R (1969) *Helping and Human Relations*, (2 volumes). New York: Holt, Rinehart and Winston.
CARRIER, Joseph M (1977) 'Sex — Role Preference as an Explanatory Variable in Homosexual Behaviour', *Archives of Sexual Behaviour*, 6 (1), 53–65.
CARRIGAN, Tim; CONNELL, Bob and **LEE, John** (1987) 'Toward a New Sociology of Masculinity'. In: Brod, Harry (Ed.) *The Making of Masculinities. The New Men's Studies.* Boston: Allen & Unwin, 63–100.
CASS, Vivienne C (1979) 'Homosexual Identity Formation', *Journal of Homosexuality*, 4 (3), 219–35.
CASS, Vivienne C (1984) 'Homosexual Identity: A Concept in Need of Definition', *Journal of Homosexuality*, 9 (2/3), 105–26.
CATHOLIC INSTITUTE FOR INTERNATIONAL RELATIONS (1988) *Now Everyone is Afraid. The Changing Face of Policing in South Africa.* London: Catholic Institute for International Relations (anonymous).
CENTRES FOR DISEASE CONTROL (ATLANTA, USA) (1982) *Morbidity and Mortality Weekly Report*, 131 (19), 249.
CHAPMAN, Beate E and **BRANNOCK, Joanne** (1987) 'Proposed Model of Lesbian Identity Development: An Empirical Examination', *Journal of Homosexuality*, 14 (3), 69–80.
CHESCHEIR, Martha (1985) 'Some Implications of Winnicott's Concept for Clinical Practice', *Clinical Social Work Journal*, 13 (3), 218–33.
CHESEBRO, James W (Ed.) (1981) *Gayspeak. Gay Male and Lesbian Communication.* New York: Pilgrim Press.
CLARK, Don (1977) *Loving Someone Gay.* Millbrae, California: Celestial Arts.
COHEN, L; CLAIBORN, W and **SPECTOR, G A** (1983) *Crisis Intervention*, Second Edition. New York: Human Sciences Press.
COLEMAN, Eli (1982) 'Developmental Stages of the Coming Out Process'. In: Gonsiorek, John C (Ed.) *Homosexuality and Psychotherapy: A Practitioner's Handbook of Affirmative Models.* New York: Haworth Press, 31–43.
COLEMAN, Eli (1987) 'Assessment of Sexual Orientation', *Journal of Homosexuality*, 14 (1/2), 9–24.

COLGAN, Phillip (1987) 'Treatment of Identity and Intimacy Issues in Gay Males', *Journal of Homosexuality*, 14 (1/2), 101–23.

COTTON, Wayne L (1972) 'Role-Playing Substitutions Among Homosexuals', *The Journal of Sex Research*, 8 (4), 310–23.

CRAMER, David W and **ROACH, Arthur J** (1988) 'Coming Out to Mom and Dad: A Study of Gay Males and Their Relationships with Their Parents', *Journal of Homosexuality*, 15 (3/4), 79–91.

CROWLEY, John W (1987) 'Howells, Stoddard and Male Homosocial Attachment in Victorian America'. In: Brod, Harry (Ed.) *The Making of Masculinities. The New Men's Studies.* Boston: Allen and Unwin, 301–24.

CUMMING, John and **CUMMING, Elaine** (1964) *Ego and Milieu. Theory and Practice of Environmental Therapy.* Britain: Tavistock Publications.

DAHER, Douglas (1977) 'Sexual Identity Confusion in Late Adolescence: Therapy and Values', *Psychotherapy: Theory, Research and Practice*, 14 (1), 12–17.

DALDIN, Herman (1988) 'The Fate of the Sexually Abused Child', *Clinical Social Work Journal*, 16 (1), 22–31.

DANK, Barry M (1971) 'Coming Out in the Gay World', *Psychiatry*, 34 (2), 180–97.

DARBONNE, Allen R (1967) 'Crisis: A Review of Theory, Practice and Research', *Psychotherapy: Theory, Research and Practice*, 4 (2), 49–56.

DARDICK, Larry and **GRADY, Kathleen** (1980) 'Openness between Gay Persons and Health Professionals', *Annals of Internal Medicine*, 93 (part 1), 115–19.

DAY, Peter R (1981) *Social Work and Social Control.* London and New York: Tavistock.

DE CECCO, John P (1981) 'Definition and Meaning of Sexual Orientation', *Journal of Homosexuality*, 6 (4), 51–67.

DE CECCO, John P and **SHIVELY, Michael G** (Eds.) (1984a) *Bisexual and Homosexual Identities: Critical Theoretical Issues.* New York: Haworth Press.

DE CECCO, John P and **SHIVELY, Michael G** (1984b) 'From Sexual Identity to Sexual Relationships: A Contextual Shift'. In: De Cecco, John P and Shively, Michael G (Eds.) *Bisexual and Homosexual Identities: Critical Theoretical Issues.* New York: Haworth Press, 1–26.

DECRESCENZO, Teresa A (1984) 'Homophobia: A Study of the Attitudes of Mental Health Professionals toward Homosexuality', *Journal of Social Work and Human Sexuality*, 2 (2/4), 115–36.

DE COCK, Kevin (1984) 'AIDS: An Old Disease from Africa', *British Medical Journal*, 289, 301–8.

DE GRUCHY, John W (1985) 'Theologies in conflict: The South African Debate'. In: Villa-Vicenceo, Charles and De Gruchy, John W (Eds.) *Resistance and Hope. South African Essays in Honour of Beyers Naude.* Cape Town and Johannesburg: David Philip, 85–97.

DE MONTEFLORES, Carmen and **SCHULTZ, Stephen** (1978) 'Coming Out: Similarities and Differences for Lesbians and Gay Men', *Journal of*

Social Issues, 34 (3), 59–72.
DENNENY, Michael; ORTLEB, Charles and **STEELE, Thomas** (1984) *The View from Christopher Street.* London: Hogarth Press.
DEUCHAR, N (1984) 'AIDS in New York City with Particular Reference to the Psycho-Social Aspects', *British Journal of Psychiatry,* 145, 612–19.
DILLEY, James W; OCHITILL, Herbert N; PERL, Mark and **VOLBERDING, Paul A** (1985) 'Findings in Psychiatric Consultations with Patients with Acquired Immune Deficiency Syndrome', *American Journal of Psychiatry,* 142 (1), 82–5.
DIXON, Samuel L (1979) *Working with People in Crisis. Theory and Practice.* St Louis: C V Mosby.
DUGGAN, H A (1984) *Crisis Intervention: Helping Individuals at Risk.* Lexington, Massachusetts: Lexington Books.
DUNBAR MOODIE, T (with Vivienne Ndatshe and British Sibuyi) (1988) 'Migrancy and Male Sexuality on the South African Gold Mines', *Journal of Southern African Studies,* 14 (2), 228–56.
DUNKEL, Joan and **HATFIELD, Shellie** (1986) 'Countertransference Issues in Working with Persons with AIDS', *Social Work. National Association of Social Workers,* 31 (2), 114–17.
EHRLICH, Larry G (1981) 'The Pathogenic Secret'. In: Chesebro, James W (Ed.) *Gayspeak: Gay Male and Lesbian Communication.* New York: The Pilgrim Press, 130–41.
ELPHIS, Christopher (1987) *Sexuality and Birth Control in Community Work,* Second Edition. London and New York: Tavistock Publications.
ERIKSON, E H (1956) 'The Problem of Ego Identity', *Journal of American Psychoanalytic Association,* 4, 56–121.
ERIKSON, E H (1959) 'Identity and the Life Cycle', *Psychological Issues,* (1), 1–17.
ERIKSON, Erik H (1963) *Childhood and Society,* Second Edition. New York: W W Norton.
ERIKSON, E H (1968) *Identity: Youth and Crisis.* London: Faber and Faber.
EWING, Charles P (1978) *Crisis Intervention as Psychotherapy.* New York: Oxford University Press.
EXIT (1984) No. 20, June/July.
EXIT (1987) No. 17, February/March; No. 19, May/June; No. 20, June/July; No. 22, August; No. 23, September, and No. 25, November.
EXIT (1988) No. 30, August; No. 32, October.
FAULSTICH, Michael E (1987) 'Psychiatric Aspects of AIDS', *American Journal of Psychiatry,* 144, 551–6.
FERRARA, Anthony J (1984) 'My Personal Experience with AIDS', *American Psychologist,* November, 1 285–7.
FERRINHO, H M (1981) *Towards a Theory of Community Social Work.* Cape Town: Juta and Company.
FINCH, Briony Jean (1973) 'Male and Female Homosexuality: A Comparison of Secretiveness, Promiscuity, Masculine/Feminine Role-playing, Self-concept and Self-acceptance'. Unpublished Honours dis-

sertation (psychology), University of the Witwatersrand.
FISCHER P (1972) *The Gay Mystique: The Myth and Reality of Male Homosexuality.* New York: Stein and Day.
FLAVIN, Daniel K; FRANKLIN, John E and **FRANCES, Richard J** (1986) 'The Acquired Immune Deficiency Syndrome (AIDS) and Suicidal Behaviour in Alcohol-Dependent Homosexual Men', *American Journal of Psychiatry*, 143, 1 440–2.
FLISHER, Alan John (1981) 'The Development, Implementation and Evaluation of a Training Programme in Rape Crisis Intervention for Lay Therapists: A Community Psychology Approach'. Unpublished Master's thesis (clinical psychology), University of Cape Town.
FLISHER, Alan J and **ISAACS, Gordon M** (1987) 'The Evaluation of a Training Programme in Rape Crisis Intervention for Lay Therapists', *South African Journal of Psychology*, 17 (2), 40–6.
FONE, Byrne R S (1980) *The Gay Experience: Fiction and Non-Fiction from the Homosexual Experience.* New York: AMS Press.
FORD, D H and **URBAN, H B** (1963) *Systems of Psychotherapy.* New York: John Wiley and Sons.
FORD, C S and **BEACH, F A** (1970) *Patterns of Sexual Behaviour.* New York: Harper and Row. (Originally published in 1951.)
FOUCAULT, Michael (1976) *The History of Sexuality: An Introduction.* Middlesex, England: Penguin Books. (Translated from the French by Robert Hurley.)
FOWLER, H W and **FOWLER, F G** (1951) *The Concise Oxford Dictionary of Current English*, Fourth Edition. Oxford: Clarendon Press.
FRANCHER, J Scott and **HENKIN, Janet** (1973) 'The Menopausal Queen: Adjustment to Aging and the Male Homosexual', *American Journal of Orthopsychiatry*, 43 (4), 670–4.
FREEDMAN, Alfred M; KAPLAN, Harold J and **SADOCK, Benjamin** (1976) *Modern Synopsis of Comprehensive Textbook of Psychiatry/II.* Baltimore: Williams and Wilkins.
FREUD, Sigmund (1977) *On Sexuality. Three Essays on the Theory of Sexuality*, Vol. 7. Compiled and edited by Angela Richards. Middlesex, England: Penguin Books.
FREUND, K and **BLANCHARD, B** (1983) 'Is the Distant Relationship of Fathers and Homosexual Sons Related to the Sons' Erotic Preference for Male Partners, or to the Sons' Atypical Gender Identity, or to Both?' In: Ross, Michael E (Ed.) *Homosexuality and Social Sex Roles.* New York: Haworth Press, 7–25.
FRIEND, Richard A (1987) 'The Individual and Social Psychology of Aging: Clinical Implications for Lesbians and Gay Men', *Journal of Homosexuality*, 14 (1/2), 307–31.
FROSCH, John (1981) 'The Role of Unconscious Homosexuality in the Paranoid Constellation', *Psychoanalytic Quarterly.* L, 587–611.
GADPAILLE, Warren J (1980) 'Cross-Species and Cross-Cultural Contri-

butions to Understanding Homosexual Activity', *Archives of General Psychiatry*, 37 (3), 349–56.

GAGNON, J H and SIMON, W (1967) 'Femininity in the Lesbian Community', *Social Problems*, 15, 212–21.

GAGNON, J H and SIMON, W (1974) *Sexual Conduct: The Social Sources of Human Sexuality*. London: Hutchinson.

GALLO, R; SALAHUDDIN, S; POPOVIC, M and Associates (1984) 'Frequent Detection and Isolation of Cytopathic Retroviruses (HTLVIII) From Patients with AIDS and at Risk for AIDS', *Science*, 224, 500–3.

GAMBRILL, Eileen (1983) *Casework: A Competency-Based Approach*. New Jersey: Prentice-Hall.

GASA 60-10 (1988) *Counselling Service Annual Report*. (Compiled by John Pegge, Director.) Cape Town: GASA 60-10.

GAY LEFT COLLECTIVE (Eds.) (1980) *Homosexuality: Power and Politics*. London and New York: Allison and Busby Limited.

GEARHART, Sally Miller (1981) 'Gay Civil Rights and the Roots of Oppression'. In: Chesebro, James W (Ed.) *Gayspeak: Gay Male and Lesbian Communication*. New York: The Pilgrim Press, 275–85.

GEDDES, Donald Porter (1954) *An Analysis of the Kinsey Report on Sexual Behaviour in the Human Male and Female*. New York: Mentor Books, The New American Library of World Literature.

GEORGE, Kenneth D and BEHRENDT, Andrew E (1987) 'Therapy for Male Couples Experiencing Relationship Problems and Sexual Problems', *Journal of Homosexuality*, 14 (1/2), 77–88.

GERMAIN, Carel B and GITTERMAN, Alex (1980) *The Life Model of Social Work Practice*. New York: Columbia University Press.

GERSHMAN, Harry (1983) 'The Stress of Coming Out', *American Journal of Psychoanalysis*, 43 (2), 129–38.

GILLIS, L S (1986) *Guidelines in Psychiatry*, Third Edition. Cape Town: Juta and Co. Ltd.

GLANZ, L E (1987) *Attitudes of White South Africans Toward Certain Legal Rights of Homosexuals*. Memorandum submitted to the President's Council. Pretoria: Human Science Research Council.

GOCHROS, Harvey L (1984) 'Teaching Social Workers to Meet the Needs of the Homosexuality Oriented', *Journal of Social Work and Human Sexuality*, 2 (2/3) 137–48.

GOCHROS, H and SCHULTZ, E (1972) *Human Sexuality and Social Work*. New York: Associated Press.

GOFFMAN, Erving (1963) *Stigma. Notes on the Management of Spoiled Identity*. Middlesex, England: Penguin Books.

GOLAN, Naomi (1978) *Treatment in Crisis Situations*. New York: Free Press.

GOLAN, Naomi (1981) *Passing through Transitions. A Guide for Practitioners*. New York: Free Press.

GOLAN, Naomi and VASHITZ, Batya (1974) 'Social Services in a War Emergency', *Social Service Review*, 48, 422–7.

GOLDBERG, Richard L (1984) 'Heterosexual Panic', *American Journal of Psychiatry*, 44 (2), 209–11.

GONEN, Jay Y (1971) 'Negative Identity in Homosexuals', *Psychoanalytic Review*, 58 (3), 345–52.

GONG, Victor (Ed.) (1985) *Understanding AIDS. A Comprehensive Guide*. Cambridge: Cambridge University Press.

GONG, Victor (1985) 'Preface — Update on AIDS Research'. In: Gong, Victor (Ed.) *Understanding AIDS — A Comprehensive Guide*. Cambridge, London: Cambridge University Press, xv–xxii.

GONSIOREK, John C (Ed.) (1982) *Homosexuality and Psychotherapy. A Practitioner's Handbook of Affirmative Models*. New York: Haworth Press.

GONSIOREK, John C (1982) 'The Use of Diagnostic Concepts in Working with Gay and Lesbian Populations'. In: Gonsiorek, John C (Ed.) *Homosexuality and Psychotherapy. A Practitioner's Handbook of Affirmative Models*. New York: Haworth Press, 9–20.

GOODE, E (1981) 'Comments on the Homosexual Role', *Journal of Sex Research*, 17, 54–65.

GOULD R (1972) 'The Phases of Adult Life: A Study in Developmental Psychology', *American Journal of Psychiatry*, 129 (5), 521–31.

GOULDEN, Terry; **TODD, Peter**; **HAY, Robert** and **DYKES, Jim** (1984) 'AIDS and Community Supportive Services. Understanding and Management of Psychological Needs', *The Medical Journal of Australia*, October, 582–6.

GRAMICK, Jeannine (1983) 'Homophobia: A New Challenge', *Social Work. Journal of the National Association of Social Workers*, 28 (2), 137–41.

GREEN, Richard (Ed.) (1975) *Human Sexuality. A Health Practitioner's Text*. Baltimore: Williams and Wilkins.

GREENBERG, David F and **BYSTRYN, Marcia H** (1984) 'Capitalism, Bureaucracy and Male Homosexuality', *Contemporary Crises*, 1 (1), 33–56.

GREENBERG, Jerrold S (1976) 'The Effects of a Homophile Organisation on Self-Esteem and Alienation of Its Members', *Journal of Homosexuality*, 1 (3), 313–18.

GRINNELL, Richard M (Jun.) (1981) *Social Work Research and Evaluation*. Itasca, Illinois: F E Peacock Publishers.

HABERMAS, J (1979) *Communication and the Evolution of Society*. Boston: Beacon Press.

HALL, Martin (1986) 'Resistance and Rebellion in Greater Cape Town, 1985'. Unpublished monograph. Cape Town: Centre for African Studies, University of Cape Town.

HALPERN, Howard A (1973) 'Crisis Theory: A Definitional Study', *Community Mental Health Journal*, 9 (4), 342–9.

HAMMERSMITH, Sue Kiefer and **WEINBERG, Martin S** (1973) 'Homosexual Identity: Commitment, Adjustment and Significant Others', *Sociometry*, 36 (1), 56–79.

HAMMERSMITH, Sue Kiefer (1987) 'A Sociological Approach to Counselling Homosexual Clients and Their Families', *Journal of Homosexuality*, 14 (1/2), 173–89.

HAMMETT, Theodore M (1986) *AIDS in Prisons and Jails: Issues and Options*. National Institute of Justice: U S Dept of Justice.

HAROWSKI, Kathy J (1987) 'The Worried Well: Maximising Coping in the Face of AIDS', *Journal of Homosexuality*, 14 (1/2), 292–306.

HARRY, Joseph and **DEVALL, William** (1978) 'Age and Sexual Culture among Homosexually-Oriented Males', *Archives of Sexual Behaviour*, 7 (3), 199–209.

HART, John (1979) *Social Work and Sexual Conduct*. London: Routledge and Kegan Paul.

HART, John (1984) 'Therapeutic Implications of Viewing Sexual Identity in Terms of Essentialist and Constructionist Theories', *Journal of Homosexuality*, 9 (4), 39–51.

HARTMAN, A (1978) 'Diagrammatic Assessment of Family Relationships', *Social Casework*, October, 465–76.

HASTINGS, G B; LEATHER, D S and **SCOTT, S** (1987) 'AIDS Publicity: Some Experiences from Scotland', *British Medical Journal*, 294, 48–50.

HAUSER, R (1962) *The Homosexual Society*. London: The Bodley Head.

HAYES, Joseph J (1981a) '*Lesbians, Gay Men, and their "Languages"*'. In: Chesebro, James W (Ed.) *Gayspeak: Gay Male and Lesbian Communication*. New York: The Pilgrim Press, 28–42.

HAYES, Joseph J (1981b) 'Gayspeak'. In: Chesebro, James W (Ed.) *Gayspeak: Gay Male and Lesbian Communication*. New York: The Pilgrim Press, 45–67.

HEARN, Jeff and **PARKIN, Wendy** (1987) '*Sex' at 'Work'. The Power and Paradox of Organisation Sexuality*. Sussex: Wheatsheaf Book.

HEBDIGE, Dick (1979) *Subculture: The Meaning of Style*. London and New York: Methuen.

HELLMANN, Ronald E; GREEN, Richard; GRAY, James L and **WILLIAMS, Katherine** (1981) 'Childhood Sexual Identity, Childhood Religiosity, and "Homophobia" as Influences in the Development of Transsexualism, Homosexuality, and Heterosexuality', *Archives of General Psychiatry*, 38, 910–15.

HELM, Brunhilde (1973) 'Deviant or Variant? Some Sociological Perspectives on Homosexuality and its Subculture'. In: *ASSA: Sociology Southern Africa*. Papers from the First [sic] Congress of the Association for Sociologists in Southern Africa. Durban: The Association.

HELQUIST, Michael (1987) 'Your HIV Status: What Does It Mean If You Test Positive?', *The Advocate*, July, 42–7.

HENCKEN, Joel D (1984) 'Conceptualisation of Homosexual Behaviour which Preclude Homosexual Self-Labeling', *Journal of Homosexuality*, 4 (9), 53–63.

HENLEY, Clark (1982) *The Butch Manual. The Current Drag and How To Do It*. New York City: The Sea Horse Press.

HEPWORTH, Dean H and **LARSEN, Jo Ann** (1986) *Direct Social Work Practice. Theory and Skills*, Second Edition. Chicago, Illinois: The Dorsey Press.

HETRICK, Emery S and **MARTIN, A Damien** (1987) 'Developmental Issues and Their Resolution for Gay and Lesbian Adolescents', *Journal of Homosexuality*, 14 (1/2), 25–43.

HIRSCHOWITZ, Ralph G (1972) 'Crisis Theory'. Unpublished paper presented to National Multi-Professional Conference in Johannesburg, June 1972.

HODGES, Andrew and **HUTTER, David** (1974) *With Downcast Gays. Aspects of Homosexual Self-Oppression*. London: Pomegranate Press.

HOFF, Lee Ann (1978) *People in Crisis. Understanding and Helping*. California: Addison-Wesley Publishing Company.

HOFFMAN, Martin (1968) *The Gay World. Male Homosexuality and the Social Creation of Evil*. New York: Bantam Books.

HOFFMANN, Richard J (1984) 'Vices, Gods and Virtues: Cosmology as a Mediating Factor in Attitudes Toward Male Homosexuality'. In: De Cecco, John P and Shively, Michael G (Eds.) *Bisexual and Homosexual Identities: Critical Theoretical Issues*. New York: Haworth Press, 27–44.

HOLLAND, Jimmie C and **TROSS, Susan** (1987) 'Psychosocial Considerations in the Therapy of Epidemic Kaposi's Sarcoma', *Seminars in Oncology*, 14 (2), Supp. 3 (June), 48–53.

HOLLIS, Francis and **WOODS, Mary E** (1981) *Casework. A Psychosocial Therapy*, Third Edition. New York: Random House.

HOOKER, Evelyn (1965) 'A Preliminary Analysis of the Group Behaviour of Homosexuals', *Journal of Psychiatry*, 42, 217–25.

HORWITZ, T L (1981) 'Attitudes of White and Coloured University Students in South Africa Towards Male Homosexuality'. Unpublished Master's dissertation (psychology), University of South Africa.

HOULT, Thomas Ford (1984) 'Human Sexuality in Biological Perspective: Theoretical and Methodological Considerations'. In: De Cecco, John P and Shively, Michael G (Eds.) *Bisexual and Homosexual Identities: Critical Theoretical Issues*. New York: Haworth Press, 137–55.

HUMPHREYS, Laud (1972) *Out of the Closets: The Sociology of Homosexual Liberation*. Englewood Cliffs, New Jersey: Prentice-Hall.

HUMPHREYS, L (1979) 'Exodus and Identity: The Emerging Gay Culture'. In: Levine, M P (Ed.) *Gay Men: The Sociology of Male Homosexuality*, New York: Harper and Row.

HUNT, P M A (1982) *South African Criminal Law and Procedure*, (Vol. 2), Second Edition. Cape Town: Juta and Co.

IJSSELMUIDEN, C B; STEINBERG, M H and Associates (1988) 'AIDS and South Africa — Towards a Comprehensive Strategy, Part 1. The Worldwide Experience', *South African Medical Journal*, 73, 455–60.

ISAAC, S and **MICHAEL, W D** (1977) *Handbook in Research and Evaluation: For Behavioural Sciences*. San Diego, California: Edits Publishers.

ISAACS, G (1979a) 'Working with the Male Homosexual Client: Some Major Theoretical and Clinical Assumptions'. Part 1 of an unpublished Master's dissertation (social work), University of Cape Town.

ISAACS, G (1979b) 'Crisis Intervention as a Form of Therapy for Persons with Homosexual Concerns: An Experimental Study'. Part 2 of an unpublished Master's dissertation (social work), University of Cape Town.

ISAACS, G M (1985) 'Gay Oppression in South Africa: Social and Psychological Implications'. Unpublished paper presented at First National Conference on Homosexuality. Gay Association of South Africa. Johannesburg, 31 May–2 June.

ISAACS, G M (1987a) 'Crisis Psychotherapy with Persons Experiencing the AIDS Related Complex', *Crisis Intervention*, 4 (4), 115–21.

ISAACS, G M (1987b) 'AIDS: The Challenge of Social Work', *Maatskaplike Werk/Social Work*, 23 (3), 151–2.

ISAACS, G (1988) 'Counselling Methodology within the Framework of Human Sexuality', *Medical Sex Journal of South Africa* (Supplement of S A Family Practice), 9 (6), 33–4.

ISAACS, G and MILLER, D (1985) 'AIDS — Its Implications for South African Homosexuals and the Mediating Role of the Medical Practitioner', *South African Medical Journal*, 68, 327–30.

ISAACS, Gordon and ZIMBLER, Allen (1984) *General Crisis Intervention: Guidelines and Considerations*. Training manual presented to volunteers and professionals. Johannesburg Radio 702 Crisis Centre.

JACOBSON, G; STRICKLER, M and MORLEY, W E (1968) 'Generic and Individual Approaches to Crisis Intervention', *American Journal of Public Health*, 58, 339–42.

JANDT, Fred E and DARSEY, James (1981) 'Coming Out as a Communicative Process'. In: Chesebro, James W (Ed.) *Gayspeak: Gay Male and Lesbian Communication*. New York: The Pilgrim Press, 12–27.

JAY, Karla and YOUNG, Allen (1972) *Out of the Closets: Voices of Gay Liberation*. New York: World Publishing.

JAY, Karla and YOUNG, Allen (1978) *Lavender Culture*. New York: Jove Publications.

JAY, Karla and YOUNG, Allen (1979) *The Gay Report. Lesbians and Gay Men Speak Out about Sexual Experiences and Lifestyles*. New York: Summit Books.

JOFFE, Hugh Ivan (1980) 'An Attempt to Minimize the Adjustment Reaction of Aged Home Entrants in the Greater Cape Town Area'. Unpublished Ph.D. thesis (psychology), University of Cape Town.

JOHNSON, Edward and HO, Peter (1985) 'The Elusive Etiology — Possible Causes and Pathogenesis'. In: Gong, Victor (Ed.) *Understanding AIDS. A Comprehensive Guide*. Cambridge: Cambridge University Press, 77–89.

JOHNSON, H (1972) 'Ideology and the Social System', *International Encyclopaedia of the Social Sciences*, 7. New York: Macmillan Company and Free Press, 76–85.

JOSEPH, Jill G; EMMONS, Carol-Ann; KESSLER, Ronald C; WORTMAN, Camille B; O'BRIEN, Kerth; HOCTER, William T and SCHAEFER, Catherine (1984) 'Coping with the Threat of AIDS: An Approach to Psychosocial Assessment', *American Psychologist*, 39 (11), 1 297–1 302.

JOUBERT, Dian (1974) *Tot Dieselfde Geslag: Debat oor Homoseksualiteit in 1968* (Brandpunte 7). Cape Town: Tafelberg.

JOUBERT, D (1985) 'Homoseksualiteit', *Maatskaplike Werk/Social Work*, 21 (1), 36–43.

JOUBERT, Dian; CONRADIE, Ernst L; LOUW, Daniel J and HURST, Lewis A (1980) *Perspektiewe op Homosexualiteit*. Durbanville: Uitgewery Boschendal.

KADUSHIN, Alfred (1983) *The Social Work Interview*, Second Edition. New York: Columbia University Press.

KAHN, Raphael (1978) 'Therapist Variables in Crisis Intervention Therapy'. Unpublished Master's dissertation (clinical psychology), University of the Witwatersrand.

KAPLAN, D M (1968) 'Observations on Crisis Theory and Practice', *Social Casework: The Journal of Contemporary Social Work*, 49, 151–5.

KARR, Rodney G (1981) 'Homosexuality and the Male Role'. In Chesebro, James W (Ed.) *Gayspeak: Gay Male and Lesbian Communication*. New York: Pilgrim Press, 3–11.

KATZ, Jonathan (1976) *Gay American History. Lesbians and Gay Men in the USA*. New York: Thomas Y Cromwell Company.

KATZ, Jonathan Ned (1983) *Gay/Lesbian Almanac: A New Documentary*. New York: Harper and Row.

KENNETH, George D and BEHRENDT, Andrew E (1987) 'Therapy for Male Couples Experiencing Relationship Problems and Sexual Problems', *Journal of Homosexuality*, 14 (1/2), 77–88.

KING, Dave (1984) 'Condition, Orientation, Role or False Consciousness? Models of Homosexuality and Transsexualism', *The Sociological Review*, 32 (1), 38–56.

KINGSLEY, Lawrence; KASLOW, Richard and MULTICENTRE AIDS COHORT STUDY (1987) 'Risk Factors for Seroconversion to Human Immunodeficiency Virus Among Male Homosexuals', *The Lancet*, 14 February, 345–9.

KINSEY, Alfred C; POMEROY, Wardells and MARTIN, Clyde E (1948) *Sexual Behaviour in the Human Male*. London: Saunders.

KIVOWITZ, Alexandra L (1988) 'Being Missed and Missing: An Interactional Element of Separation/Individuation', *Clinical Social Work Journal*, 16 (3), 261–9.

KLEIN, F; SEPEKOFF, B and WOLF, T J (1985) 'Sexual Orientation: A Multi-Variate Dynamic Process', *Journal of Homosexuality*, 11 (12), 3 549.

KLEIN, M (1959) 'Our Adult World and Its Roots in Infancy', *Human Relations*, XII (4), 292–303.

KNOBEL, G J (1986) 'AIDS in Dental Practice: Hazards, Infection Control,

Medico-legal and Ethical Aspects', *Journal of Forensic Odonto-stomatology*, 4 (1), 15–27.

KNUTSON, Donald C (1979) 'Job Security for Gays: Legal Aspects'. In: Berzon, Betty and Leighton, Robert (Eds.) *Positively Gay*. Millbrae, California: Celestial Arts, 171–87.

KOTZE, C G (1975) 'Die Aard en Betekenis van die Homoseksuele Subkultuur', *The South African Journal of Sociology*, 12, 81–8.

KREUGER, David W (1983) 'Childhood Parent Loss: Developmental Impact and Adult Psychology', *American Journal of Psychotherapy*, Vol. XXXVII (4), October, 582–92.

KRIEGLER, R and STAFFORD, R (1988) 'S v Matsemela en 'n ander (Transvaalse Provinsiale Afdeling)', *South African Law Reports*, (2), 254–8.

KRONEMEYER, Robert (1980) *Overcoming Homosexuality*. New York: Macmillan Publishing Company.

KUBLER-ROSS, E (1969) *On Death and Dying*. New York: Macmillan.

LA TORRE, Ronald and WENDENBERG, Kristina (1983) 'Psychological Characteristics of Bisexual, Heterosexual and Homosexual Women'. In: Ross, Michael W (Ed.) *Homosexuality and Social Sex Roles*. New York: The Haworth Press, 87–97.

LAURITSEN, John and THORSTAD, David (1974) *The Early Homosexual Rights Movement (1864–1935)*. New York: Times Change Press.

LAYON, J; WARZYNSKI, M and IDRIS, A (1986) 'Acquired Immunodeficiency Syndrome in the United States: A Selective Review', *Critical Care Medicine*, 14 (9), 819–27.

LEE, Alfred McClung (1966) *Multivalent Man*. New York: George Braziller.

LEE, John Alan (1977) 'Going Public: A Study in the Sociology of Homosexual Liberation', *Journal of Homosexuality*, 3 (1) 49–78.

LESTER, David (1975) 'The Relationship between Paranoid Delusions and Homosexuality', *Archives of Sexual Behaviour*, 4 (3), 285–93.

LEVINE, Martin P (1977) 'Gay Ghetto', *Journal of Homosexuality*, 4 (4), 363–78.

LIDDICOAT, Renee (1961)'A Study of Non-Institutionalised Homosexuals', *Journal of National Institute for Personnel Research*, 8, 217–49.

LINDEMANN, E (1944) 'Symptomatology and Management of Acute Grief', *American Journal of Psychiatry*, 101 (2), 141–8. Reprinted in Parad, H J (Ed.) *Crisis Intervention: Selected Readings*. New York: Family Service Association of America, 1965, 7–21.

LITTLEWOOD, Roland and LIPSEDGE, Maurice (1982) *Aliens and Alienists. Ethnic Minorities and Psychiatry*. Middlesex, England: Penguin Books.

LOPEZ, Diego J and GETZEL, George S (1984) 'Helping Gay AIDS Patients in Crisis', *Social Casework: The Journal of Contemporary Social Work*, September, 387–94.

LOPEZ, Diego and GETZEL, George S (1987) 'Strategies for Volunteers

Caring for Persons with AIDS', *Social Casework: The Journal of Contemporary Social Work*, January, 47–53.

LORAINE, J A (1974) *Understanding Homosexuality. Its Biological and Psychological Bases.* Lancaster, England: Medical and Technical Publishing Company.

LOWE, Gary R (1988) 'South African Social Work and the Norm of Injustice', *Social Service Review*, 62 (1), 20–39.

LUKTON, Rosemary Creed (1974) 'Crisis Theory: Review and Critique', *Social Service Review*, 48 (3), 384–402.

MAHLER, M (1971) 'A Study of the Separation-Individuation Process and Its Possible Application to Borderline Phenomena in the Psychoanalytic Situation', *The Psychoanalytic Study of the Child*, 26, 403–24.

MAHLER, M (1974) 'Symbiosis and Individuation: The Psychological Birth of the Human Infant', *The Psychoanalytic Study of the Child*, 29, 89–106.

MALAN, D H (1976) *The Frontier of Brief Psychotherapy.* New York: Plenum Press.

MALAN, D H (1979) *Individual Psychotherapy and the Science of Psychodynamics.* London: Butterworths.

MALYON, Alan K (1982a) 'Psychotherapeutic Implications of Internalised Homophobia in Gay Men'. In: Gonsiorek, John C (Ed.) *Homosexuality and Psychotherapy: A Practitioner's Handbook of Affirmative Models.* New York: Haworth Press, 59–70.

MALYON, Alan K (1982b) 'Biphasic Aspects of Homosexual Identity Formation', *Psychotherapy: Theory, Research and Practice*, 19 (3), 335–40.

MANN, James (1973) *Time-Limited Psychotherapy.* Cambridge: Harvard University Press.

MARGO, Glen (1976) 'Social Stereotyping. A Study of Discriminatory Behaviour in the Delivery of Health Care Services to Sexual Minorities'. Unpublished D.P.H. thesis, Berkley University, San Francisco.

MAROTTA, Toby (1981) *The Politics of Homosexuality.* Boston: Houghton Mifflin.

MARSH, Michael; GONG, Victor and **SHINDLER, Daniel** (1985) 'Questions and Answers about AIDS'. In: Gong, Victor, *Understands AIDS. A Comprehensive Guide.* Cambridge, London: Cambridge University Press, 190–8.

MARTIN, John L and **VANCE, Carol S** (1984) 'Behavioural and Psychosocial Factors in AIDS', *American Psychologist*, 39 (11), 1 303–8.

MASLOW, Abraham H (1962) *Toward a Psychology of Being.* Princeton, New Jersey: D van Nostrand Company.

MASON, Judy (1987) 'Social Work with Individuals and Families'. In: McKendrick, Brian (Ed.) *Introduction to Social Work.* Pinetown: Owen Burgess Publishers, 47–75.

MASTERS, William H and **JOHNSON, Virginia E** (1979) *Homosexuality in Perspective.* Boston: Little, Brown.

MAY, R; ANGEL, E and ELLENBERGER, H F (Eds.) (1958) *Existence: A New Dimension in Psychiatry and Psychology*. New York: Basic Books.

McDONALD, Gary J (1982) 'Individual Differences in the Coming Out Process for Gay Men: Implications for Theoretical Models', *Journal of Homosexuality*, 8 (1), 47–60.

McKENDRICK, Brian W (Ed.) (1987) *Introduction to Social Work in South Africa*. Pinetown: Owen Burgess.

McKENDRICK, Brian W (1987a) 'The Development of Social Welfare and Social Work in South Africa'. In: McKendrick, Brian W (Ed.) *Introduction to Social Work in South Africa*. Pinetown: Owen Burgess, 5–19.

McKENDRICK, Brian W (1987b) 'The South African Welfare System'. In: McKendrick, Brian W (Ed.) *Introduction to Social Work in South Africa*. Pinetown: Owen Burgess, 20–46.

McKUSICK, L; CONANT, M and COATES, T (1985a) 'The AIDS Epidemic: A Model for Intervention. Strategies for Reducing High-Risk Behaviour in Gay Men', *Sexually Transmitted Diseases*, Oct-Dec, 229233.

McKUSICK, L; HORSTMAN, W and COATES, T J (1985b) 'AIDS and Sexual Behaviour Reported by Gay Men in San Francisco', *American Journal of Public Health* 75, 493–6.

McWHIRTER, David P and MATTISON, Andrew M (1984) *The Male Couple: How Relationships Develop*. Englewood Cliffs, New Jersey: Prentice-Hall.

MEAD, George H (1934) *Mind, Self and Society*. Chicago and London: University of Chicago Press.

MEDORA, Nilufer and CHESSER, Barbara (1980) 'Brief Report: Two Studies of Hope and Faith in Coping with Crises', *Crisis Intervention*, 11 (4), 141–7.

MEHR, Joseph (1983) *Human Services: Concepts and Intervention Strategies*, Second Edition. Boston: Allyn and Bacon.

MESSING, Alice E; SCHOENBERG, Robert and STEPHENS, Roger K (1984) 'Confronting Homophobia in Health Care Settings: Guidelines for Social Work Practice', *Journal of Social Work and Human Sexuality*, 2 (2/3), 65–74.

MEYER, William S (1988) 'On the Mishandling of "Anger" in Psychotherapy', *Clinical Social Work Journal*, 16 (4) 406–17.

MIELI, Mario (1980) *Homosexuality and Liberation: Elements of a Gay Critique*. London: Gay Men's Press.

MILESKI, Maureen and BLACK, Donald J (1972) 'The Social Organisation of Homosexuality', *Urban Life and Culture*, 1, 187–202.

MILLER, David (1987) 'Counselling: ABC of AIDS', *British Medical Journal*, 294, 1 671–4.

MILLER, David and BROWN, Barrie (1988) 'Developing the Role of Clinical Psychology in the Context of AIDS', *The Psychologist. Bulletin of the British Psychological Society*, 2, 63–6.

MILLER, D; JEFFRIES, D J; WILLIE-HARMS, J R and PUNCHING, A J

(1986) 'HTLV-III: Should Testing be Routine?', *British Medical Journal*, 292, 941–3.

MILLER, Martin (1981) 'The Coming Out Crisis of Homosexuality: An Affirmation of Emergent Identity'. Unpublished Honours dissertation (psychology), University of Cape Town.

MILLER, Robert D (1978) 'Pseudohomosexuality in Male Patients with Hysterical Psychosis: A Preliminary Report', *American Journal of Psychiatry*, 135 (1), 112–13.

MILLIGAN, Don (1975) 'Homosexuality: Sexual Needs and Social Problems'. In: Baily, Roy and Brake, Mike (Eds.) *Radical Social Work*. London: Edward Arnold, 96–111.

MINTON, Henry L and McDONALD, Gary J (1984) 'Homosexual Identity Formation as a Developmental Process', *Journal of Homosexuality*, 9 (2/3), 91–104.

MINUCHIN, S (1974) *Families and Family Therapy*. Cambridge, Massachusetts: Harvard University Press.

MOKHOBO, D (1988) *Sexual Attitudes amongst Black Youth with Special Reference to AIDS*. AIDS Proceedings: Selected Papers and Task Group Reports of the AIDS Congress, Strategies for Southern Africa, May, Johannesburg, 34.

MONEY, John (1974) 'Two Names, Two Wardrobes, Two Personalities', *Journal of Homosexuality*, 1 (1), 65–70.

MONEY, John (1977) 'Bisexual, Homosexual and Heterosexual. Society, Law and Medicine', *Journal of Homosexuality*, 2, 229–33.

MONEY, John and EHRHARDT, A A (1972) *Man and Woman, Boy and Girl*. Baltimore: John Hopkins University Press.

MONEY, John and TUCKER, Patricia (1975) *Sexual Signatures. On Being a Man or a Woman*. Boston-Toronto: Little, Brown and Company.

MORIN, Stephen F and GARFINKLE, Ellen M (1981) 'Male Homophobia'. In: Chesebro, James W (Ed.) *Gayspeak: Gay Male and Lesbian Communication*. New York: The Pilgrim Press, 117–29.

MORIN, Stephen F and SCHULTZ, Stephen J (1978) 'The Gay Movement and the Rights of Children', *Journal of Social Issues*, 34 (2), 137–48.

MOSALA, Itumeleng (1985) 'African Independent Churches: A Study in Sociotheological Protest'. In: Villa-Vicencio, Charles and De Gruchy, John W (Eds.) *Resistance and Hope. South African Essays in Honour of Beyers Naude*. Cape Town and Johannesburg: David Philip, 103–11.

MOSES, A E and HAWKINS, R O (Jun.) (1982) *Counselling Lesbian Women and Gay Men: A Life Issues Approach*. St Louis, Missouri: Mosby.

MUCHMORE, Wes and HANSON, William (1982) *Coming Out Right. A Handbook for the Gay Male*. Boston: Alyson Publications.

MURPHY, Timothy F (1984) 'Freud Reconsidered: Bisexuality, Homosexuality, and Moral Judgement'. In: De Cecco, John P and Shively, Michael G (Eds.) *Bisexual and Homosexual Identities: Critical Theoretical Issues*. New York: Howarth Press, 65–78.

MURRAY, John B (1968) 'Learning in Homosexuality', *Psychological Reports*, 23, 659–62.
NG'WENO, Hilary (1987) 'AIDS in Africa: A Racist Taint', *Newsweek*, March, 4.
NICHOLS, Stuart E (1985) 'Psychosocial Reactions of Persons with the Acquired Immunodeficiency Syndrome', *Annals of Internal Medicine*, 103, 765–7.
NORMANN, Hansjurg (1983) 'Gay National Liberation. An Investigation into the Status of an Ideology in Cape Town'. Unpublished Honours dissertation (sociology research), University of Cape Town.
O'HAGAN, K P (1986) *Crisis Intervention in Social Services*. London: Macmillan.
ORNSTEIN, A (1986) 'Supportive Psychotherapy: A Contemporary View', *Clinical Social Work Journal*, 14 (1), 14–30.
OVESEY, L (1955) 'Pseudohomosexuality, the Paranoid Mechanism and Paranoia', *Psychiatry*, 18, 163–73.
OVESEY, Lionel (1969) *Homosexuality and Pseudohomosexuality*. New York: Science House.
PARAD, Howard J (Ed.) (1965) *Crisis Intervention: Selected Readings*. New York: Family Service Association of America.
PARK, Jan Carl (1981) 'Referendum Campaigns vs Gay Rights'. In: Chesebro, James W, *Gayspeak: Gay Male and Lesbian Communication*. New York: The Pilgrim Press, 286–90.
PAUL, Jay P (1984) 'The Bisexual Identity: An Idea Without Social Recognition'. In: De Cecco, John P and Shively, Michael G (Eds.) *Bisexual and Homosexual Identities: Critical Theoretical Issues*. New York: Howarth Press, 45–64.
PEARSON, G (1973) 'Social Work as the Privatised Solution', *The British Journal of Social Work*, 3 (2), 209–25.
PEGGE, J V (1988a) 'The Impact of AIDS', *Maatskaplike Werk/Social Work*, June, 24 (2), 104–15.
PEGGE, J V (1988b) *Counselling Service Annual Report*. Cape Town, GASA 60-10.
PERRY, S W and MARKOWITZ, J (1986) 'Psychiatric Intervention for AIDS — Spectrum Disorders', *Hospital and Community Psychiatry*, 37 (10), 1 001–6.
PETROPOULOS, Marina (1986) *The Facts of Life*. Cape Town: Tafelberg.
PFUHL, Erdwin H (Jun.) (1980) *The Deviance Process*. New York: D van Nostrand.
PICANO, Felice (1980) *A True Likeness. Lesbian and Gay Writing Today*. New York City: The Sea Horse.
PINNOCK, D (1986) 'Some Notes on the Blueprints of a Garrison City. Western Cape: Roots and Realities'. Unpublished monograph. Centre for African Studies, University of Cape Town.
PLECK, Joseph H; O'DONNELL, Lydia; O'DONNELL, Carl and SNAREY, John (1988) 'AIDS — Phobia, Contact with AIDS, and AIDS-Related

Job Stress in Hospital Workers', *Journal of Homosexuality*, 15 (3/4), 41-54.

PLUMMER, K (1975) *Sexual Stigma: An Interactionist Account*. London: Routledge and Kegal Paul.

PLUMMER, K (1981a) 'Going Gay: Identities, Life Cycles and Lifestyles in the Male Gay World'. In: Hart, J and Richardson, D (Eds.) *The Theory and Practice of Homosexuality*. London: Routledge and Kegan Paul.

PLUMMER, Kenneth (Ed.) (1981b) *The Making of the Modern Homosexual*. London: Hutchinson.

PORTER, Robert A (1966) 'Crisis Intervention and Social Work Models', *Community Mental Health Journal*, 2 (1), 13-21.

PRYTULA, Robert E; WELLFORD, Christopher D and DE MONBREUN, Bobby G (1979) 'Body Self-Image and Homosexuality', *Journal of Clinical Psychology*, 35 (3), 567-72.

PURYEAR, Douglas A (1979) *Helping People in Crisis*. San Francisco: Jossey-Bass Publishers.

QUADLAND, Michael and SHATTLS, William D (1987) 'AIDS, Sexuality and Sexual Control', *Journal of Homosexuality*, 14 (1/2), 277-98.

RAMOS, Reyes (1981) 'Participant Observation'. In: Grinnell, Richard M (Ed.) *Social Work Research and Evaluation*. Itasca, Illinois: F.E. Peacock Publishers, 348-60.

RAPOPORT, Lydia (1965) 'The State of Crisis: Some Theoretical Considerations'. In: Parad, Howard J (Ed.) *Crisis Intervention: Selected Readings*. New York: Family Service Association of America, 22-31.

RAPOPORT, Lydia (1970) 'Crisis Intervention as a Mode of Brief Treatment'. In: Roberts, Robert W and Nee, Robert H (Eds.) *Theories of Social Casework*, Chicago and London: University of Chicago Press, 265-312.

READ, Kenneth E (1980) *Other Voices: The Style of a Male Homosexual Tavern*. California: Chandler and Sharp.

RICHARDSON, Diane (1984) 'The Dilemma of Essentiality in Homosexual Theory'. In: De Cecco, John P and Shively, Michael G (Eds.) *Bisexual and Homosexual Identities: Critical Theoretical Issues*. New York: Haworth Press, 79-90.

RICHARDSON, D and HART, J (Eds.) (1981) *The Theory and Practice of Homosexuality*. London: Routledge and Kegan Paul.

RICHARDSON, D and HART, J (1981) 'The Development and Maintenance of a Homosexual Identity'. In: Richardson, D and Hart, J (Eds.) *The Theory and Practice of Homosexuality*. London: Routledge and Kegan Paul.

RICHMOND, Len and NOGUERA, Len (Eds.) (1979) *The New Gay Liberation Book*. Pala Alto, California: Ramparts Press.

ROBERTSON, Rosalind Nina (1986) 'Psychiatric Admission as Crisis: Attitudes of Psychiatric Hospital Staff Towards Family Members of Newly-Admitted Schizophrenic Patients'. Unpublished Master's dissertation (social work), University of Cape Town.

ROESLER, Thomas and DEISHER, Robert W (1972) 'Youthful Male Homosexuality: Homosexual Experience and the Process of Developing Homosexual Identity in Males Aged 16 to 22 Years', *Journal of the American Mental Health Association*, 219 (8), 1 018–23.

ROGERS, Cathy (1988) 'Fantasy and Implications for Therapy'. Unpublished seminar paper, Master's Programme in Clinical Social Work, School of Social Work, University of Cape Town, 1–17.

ROJEK, Chris; PEACOCK, Geraldine and COLLINS, Stewart (1988) *Social Work and Received Ideas*. London and New York: Routledge and Kegan Paul.

ROSENBAUM, Peter and BEEBE, John E (1975) *Psychiatric Treatment. Crisis/Clinic/Consultation*. New York: McGraw-Hill.

ROSS, Michael W (1980) 'Retrospective Distortion in Homosexual Research', *Archives of Sexual Behaviour*, 9 (6), 523–31.

ROSS, Michael W (1983a) *The Married Homosexual Man. A Psychological Study*. London: Routledge and Kegan Paul.

ROSS, Michael W (Ed.) (1983b) *Homosexuality and Social Sex Roles*. New York: Haworth Press.

ROSS, Michael W (1983c) 'Femininity, Masculinity and Sexual Orientation: Some Cross-cultural Comparisons', *Journal of Homosexuality*, 9 (1).

ROTTER, J B (1954) *Social Learning and Clinical Psychology*. Englewood Cliffs, New Jersey: Prentice-Hall.

ROTTER, J B (1966) 'Generalised Expectancies for Internal versus External Control of Reinforcement', *Psychological Monographs: General and Applied*, 80 (1), 1–28.

ROUSSO, Marilyn (1985) 'The Relationship Between Physical Disability and Narcissism: A Critique of the Literature', *Clinical Social Work Journal*, 13 (1), 5–17.

ROWAN, Robert L and GILLETTE, Paul J (1978) *The Gay Health Guide*. Boston: Little, Brown and Company.

RUEDA, Enrique T (1982) *The Homosexual Network. Private Lives and Public Policy*. Old Greenwich, Connecticutt: Devin Adair.

RUSK, T N (1971) 'Opportunity and Technique in Crisis Psychiatry', *Comprehensive Psychiatry*, 12, 249–63.

SAGARIN, E (1973) 'The Good Guys, the Bad Guys and the Gay Guys', *Contemporary Sociology*, 2, 3–12.

SALZBERGER-WITTENBERG, Isca (1970) *Psycho-analytic Insight and Relationships: A Kleinian Approach*. London: Routledge & Kegan Paul.

SAMUELS, Andrew (1985) *Jung and the Post-Jungians*. London: Routledge and Kegan Paul.

SARREL, Lorna J and SARREL, Philip M (1979) *Sexual Unfolding: Sexual Development and Sex Therapies in Late Adolescence*. Boston: Little, Brown and Company.

SAUL, Les J and BECK, Aaron T (1961) 'Psychodynamics of Male Homosexuality', *International Journal of Psychoanalysis*, 42, 43–8.

SAVAGE, Michael (1986) *The Cost of Apartheid.* Inaugural lecture. New Series No. 121. University of Cape Town, 27 August, 1–27.

SCHARFF, David E (1982) *The Sexual Relationship. An Object Relations View of Sex and the Family.* London and New York: Routledge.

SCHOENBERG, Bernard; CARR, Arthur C; PERETZ, David and **KUTSCHER, Austin H** (Eds.) (1970) *Loss and Grief: Psychological Management in Medical Practice.* New York: Columbia University Press.

SCHOENBERG, Robert and **GOLDBERG, Richard S** (1984) 'Introduction: Homosexuality and Social Work', *Journal of Social Work and Human Sexuality,* 2 (2/3), 1–5.

SCHURINK, W J and **SCHURINK, Evanthe M** (1982) 'The Development of the Gay Personality'. In: *The Seven Ages of Man.* Proceedings of a three day symposium presented by the Department of Nursing Science in collaboration with the Centre for Continuing Education of the University of Port Elizabeth. Bergsvlei: Promex Plastics, 109–29.

SCHURINK, W J and **SCHURINK, Evanthe M** (1983) *Some Characteristics of the Homosexual Subculture as Reflected by a South African Gay Club.* Paper presented to criminology students. Johannesburg: University of the Witwatersrand.

SCHURINK, W J (1986) *Gayness: A Sociological Perspective.* Paper presented at Mardi Gay: GASA Fourth Convention, Johannesburg.

SEALE, John (1986) *Memorandum Presented to the Select Committee, Health and Social Services.* United Kingdom: House of Commons.

SELYE, H (1974) *Stress without Distress.* New York: Lippincott.

SHARP, John (1988) 'Ethnic Group and Nation: The Apartheid Vision in South Africa'. In: Boonzaier, Emile and Sharp, John, *South African Keyword. The Uses and Abuses of Political Concepts.* Cape Town and Johannesburg: David Philip, 79–99.

SHILTS, Randy (1987) *And the Band Played On: Politics, People and the AIDS Epidemic.* Middlesex, England: Penguin Books.

SHIVELY, Michael G; JONES, Christopher and **DE CECCO, John P** (1984) 'Research on Sexual Orientation: Definitions and Methods'. In: De Cecco, John P and Shively, Michael G (Eds.) *Bisexual and Homosexual Identities: Critical Theoretical Issues.* New York: Haworth Press.

SHORR, Joseph E (1983) *Psychotherapy Through Imagery,* Second Edition. New York: Thieme-Strutten Publishers.

SIEGEL, Paul (1981) 'Androgyny, Sex-Role Rigidity and Homophobia'. In: Chesebro, James W (Ed.) *Gayspeak: Gay Male and Lesbian Communication.* New York: The Pilgrim Press, 142–52.

SIFNEOS, Peter E (1979) *Short-Term Dynamic Psychotherapy: Evaluation and Technique.* New York and London: Plenum Medical Book Company.

SILVERBERG, Robert A (1985) 'Men Confronting Death: Management Versus Self Determination', *Clinical Social Work Journal,* 13 (2).

SILVERSTEIN, Charles and **WHITE, E** (1977) *The Joy of Gay Sex. An Intimate Guide for Gay Men to the Pleasures of a Gay Lifestyle.* New York: Crown.

SINGER, J (1977) *Androgyny: Towards a New Theory of Sexuality.* London: Routledge and Kegan Paul.

SINGER, Jerome L (1981) *Daydreaming and Fantasy.* Oxford: Oxford University Press.

SIPOVA, Iva and **BRZEK, Antonin** (1983) 'Parental and Interpersonal Relationships of Transsexual and Masculine and Feminine Homosexual Men'. In: Ross, Michael W (Ed.) *Homosexuality and Social Sex Roles.* New York: Haworth Press, 75–86.

SLAIKEU, Karl (1984) *Crisis Intervention: A Handbook for Practice and Research.* Boston: Allyn and Bacon.

SMALL, L (1971) *The Briefer Psychotherapies.* New York: Brunner/Mazel.

SMITH, Jaime (1988) 'Psychopathology, Homosexuality and Homophobia', *Journal of Homosexuality*, 15 (1/2), 59–73.

SMITH, Sidney Greer (1983) 'A Comparison Among Three Measures of Social Sex-Role', *Journal of Homosexuality*, 9 (1), 99–107.

SNYMAN, J and Associates (1987) *Time-limited Intervention: A Guide for the Helping Professions.* Republic of South Africa: Human Sciences Research Council.

SOUTH AFRICA, Republic of (1950) *Population Registration Act*, No. 30 (as amended).

SOUTH AFRICA, Republic of (1950) *Group Areas Act*, No. 41 (as amended).

SOUTH AFRICA, Republic of (1957) *Immorality Act*, No. 23 (as amended).

SOUTH AFRICA, Republic of (1977) *Criminal Procedures Act*, No. 51 (as amended).

SOUTH AFRICA, Republic of (1978) *Social and Associated Workers Act*, No. 110 (as amended).

SOUTH AFRICA, Republic of (1982) *Annual Report of Criminal Offences.* Pretoria: Central Statistical Services.

SOUTH AFRICA, Republic of (1985) *Government Gazette*, 26 April.

SOUTH AFRICA, Republic of (1985) *Report of the* Ad Hoc *Committee of the President's Council on the Immorality Act (Act 23 of 1957 as amended).* Cape Town: Government Printer.

SOUTH AFRICA, Republic of (1988) *Government Gazette*, 12 February.

SPADA, James (1979) *The Spada Report: The Newest Survey of Gay Male Sexuality.* New York: New American Library.

STANFORD, John D (1981) *Spartacus International Gay Guide*, Eleventh Edition. Holland: Spartacus.

STEIN, Terry S (1988) 'Theoretical Considerations in Psychotherapy with Gay Men and Lesbians', *Journal of Homosexuality*, 15 (1/2), 75–95.

STOLLER, R J (1969) *Sex and Gender.* New York: Science House.

STOLLER, Robert J; MARMOR Judd; BIEBER, Irving; GOLD, Ronald; SOCARIDES, Charles W; GREEN, Richard and **SPITZER, Robert L**

(1973) 'A Symposium: Should Homosexuality be in the APA Nomenclature?', *American Journal of Psychiatry*, 130 (1), 1 207–17.
STRICKLIN, James Lane (1974) *The Psycho-Social Index*, Second Edition, Revised. Cape Town: with the assistance of a grant from the Editorial Board of the University of Cape Town.
STRYDOM, Katinka (1972) 'Die Psigo-Maatskaplike Aspekte van Homoseksualiteit'. Unpublished Ph.D. thesis (social work), University of Cape Town.
STULBURG, Ian and SMITH, Margaret (1988) 'Psychosocial Impact of the AIDS Epidemic on the Lives of Gay Men', *Social Work. Journal of the National Association of Social Workers*, 33 (3), 277–82.
SUNDAY TIMES, The (1986) 29 June.
SUNDAY TIMES, The (1987) 26 July.
SUPPE, Frederick (1984) 'Classifying Sexual Disorders: The Diagnostic and Statistical Manual of The American Psychiatric Association', *Journal of Homosexuality*, 9 (4), 9–28.
SURPLUS PEOPLE PROJECT, The (1983) *Forced Removal in South Africa. General Overview*. (Volume 1 of the Surplus People Project Report). Pietermaritzburg: Surplus People Project.
SUTHERLAND, S and SCHERL, D J (1970) 'Patterns of Response Among Victims of Rape', *American Journal of Orthopsychiatry*, 40, 503–11.
TAYLOR, Alan (1983) 'Conceptions of Masculinity and Femininity as a Basis for Stereotypes of Male and Female Homosexuals', *Journal of Homosexuality*, 9 (1), 37–54.
TAYLOR, C (1983) 'A Study of Voluntary Welfare Agencies Responses to the Phenomenon of Squatting by Coloured People in Cape Town'. Unpublished Master's dissertation (social work), University of Cape Town.
TEYBER, Edward B (1988) *Interpersonal Process in Psychotherapy. A Guide for Clinical Training*. Chicago, Illinois: The Dorsey Press.
THERON, Francois (1980) 'A Brief Summary of the Devlin-Hart Debate on Law and Morality'. Cape Town: Unpublished seminar paper, University of Cape Town.
THOMAS, A (1986) 'Future Economic Growth Prospects of the Western Cape: A Regional-Structural Perspective. Western Cape: Roots and Realities'. Unpublished monograph, Centre for African Studies, University of Cape Town.
THOMPSON, N L; SCHWARTZ, D M; McCANDLESS, Boyd R and EDWARDS, David A (1973) 'Parent-Child Relationships and Sexual Identity in Male and Female Homosexuals and Heterosexuals', *Journal of Consulting and Clinical Psychology*, 41 (1), 120–7.
TREVISAN, Joao S (1986) *Perverts in Paradise* (Translated by Martin Forman). London: GMP Publishers Ltd.
TRIPP, C A (1975) *The Homosexual Matrix*. New York: New American Library.
TROIDEN, R R (1979) 'Becoming Homosexual: A Model of Gay Identity Acquisition', *Psychiatry*, 42, 362–73.

TROIDEN, Richard R and **GOODE, Erich** (1980) 'Variables Related to the Acquisition of a Gay Identity', *Journal of Homosexuality*, 5 (4), 383–92.

TURNER, Victor W (1969) *The Ritual Process*. England: Penguin Books.

TYHURST, James C (1975) 'The Role of Transition States — Including Disasters — in Mental Illness'. In: Rosenbaum, Peter C and Beebe, John E, *Psychiatric Treatment. Crisis/Clinic/Consultation*. New York: McGraw-Hill Company, 17.

TYSON, Phyllis (1982) 'A Developmental Line of Gender Identity, Gender Role and Choice of Love Object', *Journal of the American Psychoanalytic Association*, 30, 61–86.

VAN ONSELEN, Charles (1982) *Studies in the Social and Economic History of the Witwatersrand 1886–1914*, Vol. 2. New Nineveh, Johannesburg: Ravan Press.

VERMOOTEN, J and **SCHABORT, J** (1987) 'S.v.C. (Witwatersrand Local Division)', *South African Law Reports*, 2, 77–82.

VIEIRA, Jeffrey (1985) 'The Haitian Link'. In: Gong, Victor (Ed.) *Understanding AIDS: A Comprehensive Guide*. Cambridge: Cambridge University Press, 90–9.

VILLA-VICENCIO, Charles (1988) *Trapped in Apartheid. A Socio-Theological History of the English-Speaking Churches*. Maryknoll, New York: Orbis Books.

VINEY, L L (1976) 'The Concept of Crisis: A Tool for Clinical Psychologists', *Bulletin of the British Psychological Society*, 29, 387–95.

WARREN, C A (1974) *Identity and Community in the Gay World*. New York: John Wiley.

WARREN, Carol A B (1977) 'Fieldwork in the Gay World: Issues in Phenomenological Research', *Journal of Social Issues*, 13 (4), 93–107.

WATSON, L (1984) 'Living with AIDS', *The Medical Journal of Australia*, 141 (9), 559–60.

WEBER, Vaughan C (1975) 'Sexual Confusion as Experienced by Adolescents and Young Adults – A Crisis Intervention Response'. Unpublished dissertation B.A. (social work), University of the Witwatersrand.

WEBSTER, Steven E (1977) 'Coming Out of the Closet: Discovering One's Homosexual Identity'. In: Zastrow, Charles and Chang, Dae H (Eds.) *The Personal Problem Solver*. Englewood Cliffs, New Jersey: Prentice-Hall, 122–34.

WEEKEND ARGUS, The (1983) 30 April.

WEEKS, Jeffrey (1977) *Coming Out. Homosexual Politics in Britain, from the Nineteenth Century to the Present*. London: Quartet Books.

WEEKS, Jeffrey (1981) 'Discourse, Desire and Sexual Deviance: Some Problems in a History of Homosexuality'. In Plummer, Kenneth (Ed.) *The Making of the Modern Homosexual*. London: Hutchinson, 76–111.

WEEKS, Jeffrey (1985) *Sexuality and Its Discontents: Meanings, Myths and Modern Sexualities*. London: Routledge and Kegan Paul.

WEIDEMANN, George H (1974) 'Homosexuality, A Survey', *Journal of the American Psychoanalytic Association*, 22 (3), 651–95.

WEINBERG, M S (1978) 'On "Doing" and "Being" Gay: Sexual Behaviour and Homosexual Male Self-Identity', *Journal of Homosexuality*, 4 (2), 143–56.

WEINBERG, Martin S and **WILLIAMS, Colin J** (1974) *Male Homosexuals: Their Problems and Adaptations*. New York: Oxford University Press.

WEINBERG, Martin S and **WILLIAMS, Colin J** (1975) 'Gay Baths and The Social Organisation of Impersonal Sex', *Social Problems*, 23, 124–36.

WEINBERG, T S (1984) 'Biology, Ideology and the Reification of Developmental Stages in the Study of Homosexual Identities', *Journal of Homosexuality*, 10 (3/4), 77–84.

WHITAM, Fredrick L (1983) 'Culturally Invariable Properties of Male Homosexuality: Tentative Conclusions from Cross-Cultural Research', *Archives of Sexual Behaviour*, 12 (3), 207–26.

WILSON, Francis and **RAMPHELE, Mamphela** (1989) *Uprooting Poverty. The South African Challenge*. Cape Town and Johannesburg: David Philip.

WINNICOTT, D W (1965) *The Maturational Processes and the Facilitating Environment*. New York: International Universities Press.

WINNICOTT, D W (1971) *Playing and Reality*. New York: Basic Books.

WIRZ, Beatrice (1988) 'The Importance of Childhood Factors as Vehicle in Dealing with Current Conflict Issues in an Adult Client'. Unpublished Master's paper (clinical social work), University of Cape Town.

WOOLFSON, L R (1980) 'Psychological Androgyny and Gender Identity in Adult Homosexual and Heterosexual Females'. Unpublished Ph.D. thesis (psychology), University of South Africa.

ZASTROW, Charles and **CHANG, Dae H** (Eds.) (1977) *The Personal Problem Solver*. Englewood Cliffs, New Jersey: Prentice-Hall.

ZIMBLER, Allen (1979) 'Critical Stage Theory: A Basic Crisis Intervention Paradigm'. In: *Proceedings, 10th International Congress for Suicide Prevention and Crisis Intervention*. Ottawa, Canada: June, 142–7.

ZIMBLER, A (1981) 'Depression and Crisis Intervention'. In: Soubrier, J P and Vedrinne, J (Eds.) *Depression and Suicide: Proceedings of the XIth Congress of the International Association for Suicide Prevention*. Paris, New York: Pergamon Press, 772–6.

ZIMBLER, A and **BARLING, J** (1975) 'Critical Stage Theory: A Possible Extension of the Crisis Sequence'. Unpublished paper presented at 27th Annual Congress of the South African Psychological Association, Johannesburg.

ZIMBLER, Allen; SOLOMON, Caryn; YOM TOV, Carmela and **GRUZD, Colin** (1985) *Conquering Corporate Stress*. Johannesburg: Divaris Stein.

Index

abandonment 161
acceptance 3, 134; *see also* self-acceptance
accidental crisis 43
accidental homosexuality 30
acknowledgement of crisis 54, 55
acknowledgement of sexual attraction 15, 20
'acting out' behaviour 9, 13, 83, 182, 187, 208
    AIDS and 123
action group 154–5
action-oriented approach 50, 52
active crisis, state of 61–2
    AIDS-related 132–4
activism 26–7, 94, 138–54, 213, 217–18; *see also* gay movement, formal; liberation; politics
activity vs passivity 60
adaptive capacities 12
adaptive mechanisms, need for 181
adjustment, acquiring levels of 134
admission of homosexuality 182
    see also coming out
adolescence 6, 7, 14–20, 180, 210
adolescent homosexuality 30
adrenalin syndrome 84
adulthood, early 17–20
Advocate 82, 104, 157
affection: casual sex and 107
affectionate behaviour, avoidance of 80
African Gay Association (AGA) 100, 157
African gays 94, 97, 100
    attitudes to whites 94
        GASA and 176
        vignette 77–8
African homosexual sub-culture 100
African National Congress 157, 217
age, attitude to 62–3, 174
age discrimination 174
age distribution of homosexuals 173
age of awareness 179–80
age of consent 153, 155
ageing, fear of 62–3
agent provocateur 150
aggression 55

AIDS 26, 62, 101, 112–37
    African 115, 117
    age, median 116
    camping linked with 109, 110
    case studies 129–30, 233–40
    causes 120–1
    coalescing effect 214
    crisis *see* AIDS crisis
    emotional responses to 116
    growth-promoting features 136, 213, 214
    heterosexual contacts 116, 119
    heterosexual reactions 119
    homosexual responses to 119–20
    homosexual sub-culture and 117–23
    impact on identity development 213–18
    information outlets 97
    liberation retarded by 162
    mutant virus 117
    opportunist infections 113, 114
    precursors 124
    pre-safer-sex/post-safer-sex 126, 127
    preventive education 136
    psychological factors 121–3
    retribution concept 162, 213
    risk factors 116, 117, 120–1, 125
    risk groups 115–16
    self-disclosure at workplace and 178–9
    stimulation of human endeavours 136
    stress factors 118–19
    symptoms 114; *see also* ARC
    terminality 136
    terminology 113–14
    unfinished business 127, 209
    Western 115
    worried well 116, 117, 118–19, 135
AIDS crisis 116
    defusion 133
    identification 127
    in homosexual context 124–5
    intervention 125–37
AIDS related complex *see* ARC
AIDS scare panic syndrome 116, 118, 119–21, 124–5, 126
alienation 6, 8, 185, 212
    service organization and 196, 197

sub-cultural 73
alternative behaviour 103
alternative culture 70, 215
alternative splinter group 95
alternatives 97, 99, 103, 109–10, 215
altruism 171, 200, 201, 211
ambivalence 11, 61, 73–5
Amnesty International 140
Amo 156
anal fixation 13, 16, 78
anal sex 82, 117, 146–7, 147, 152, 164
 AIDS and 116, 117, 123, 124
 recipient 116, 123, 124
ANC 157, 217
androgynous gays 97
androgyny 16, 32–3, 70, 215
anger 137, 208
 AIDS related 128
animus-anima issues 9, 10, 36
annihilation, sense of 137
anti-discriminatory measures 183, 218
anxiety 48–9, 184, 185, 186, 188
 AIDS related 124, 128
 excessive 28
 management 63
apartheid
 definition 143
 effects of 91–5, 142–6, 161
 effects of breaking down 92
apartness *see* separateness
apparatus 9
approval of others, loss of 48
ARC 114, 117, 135, 136
arrested development theory 1, 3
artefacts 110
artificial devices 82
assault trauma 122
assimilist concept of identity 72
attacks on homosexuality 4
attribution theory 122
Australia 217, 231
auto-stimulation 15–16, 82
avoidance 45

bars 96–8
baths 95, 123
beginning identity achievement 11, 17–19
behaviour 22
behaviour modification 19

behaviour patterns, homosexual-linked 27–34, 194
bereavement 41
bereavement overload 132
biological determinism 2
bio-psychological theories 1–2
bisexual behaviour 30–1
bisexual exploration 13, 16
bisexual three-somes 25
bisexuality 30–1
 homosexuality and 215–7
 ongoing 31
 theories 13
 transitional 216
bisexuals 103, 107, 215
 AIDS sufferers 116
blackmail, fear of 26
blame 61, 63, 208
 AIDS and 122, 128, 213
body distortion fantasies 223, 225
body fluids 118, 124, 125
body image 4, 34
bonding 12, 18, 24
 in infants 12
bookshops, gay 104–5
bopping 102
boredom 199
'burn-out' syndrome 98

cafe society 102
'camp' 105
'camp behaviour' 105
'camping' 78, 80, 81, 101 *ter*, 212
'camping it up' 102
'camping spots' 105–10; *see also* meeting places
candidiasis 135
Cape Town, Greater: gay sub-culture 89–111
caretaker
 experiences with 12, 13
 psychic separation from 12
castration anxieties 13
casual sex 107; *see also* 'camping'; transience
catharsis 51, 225
causes of homosexuality 10, 30
celibacy 119
cerebral schemata, homosexual 10
chauvinism 99

child abuse 190
child custody 153
child molesters 190
childhood, early 12–14
childhood experiences 208
childhood homosexuality 30
childhood sexual abuse 190
choice: existential conflict around 57
Church of Zion 145
churches 145
clandestine contact 107
clinical study 168
clique system 18
cloning behaviour 98, 195
closet, being in the 17, 186
   see also coming out
closet behaviour 108, 161, 178, 209, 217
closet syndrome 20, 73, 217
clubs 98–104
clustering, social 102
co-habiting 11
collective homosexuality and coming out 183
collective loss status 49
colour bar, meeting across the 93
'coloured' homosexuals 92, 93, 94, 100, 110
   GASA and 176
   gay social institutions and 99, 100
'coming out' 18–20, 22, 23, 156, 170–1, 180–92
   age 183, 210
   AIDS and 123, 132, 213
   American process 198
   bisexuality and 216
   case examples 181
   circumstances 189
   collective sexuality and 183–4
   compared with 'going public' 178
   definitions 180–2
   diagnosis and 222–3
   fantasies 182
   forced 190
   formal gay movement and 159
   gay identity acquisition and 209
   identity growth and 166–7
   in late adulthood 181–2
   need of vehicle for 197, 198
   related to helping others 201–2
   stress 23, 47
'coming out' crisis 20, 44, 46, 60, 184, 185–8
   projection 193
'coming out' patterns 161, 189, 198
'coming out' status, symbolic 141
Comment 156
commitment to homosexuality 181
common law offences 146
community, gay 22
   definition xii–xiii
   see also gay movement, formal; sub-culture, homosexual
community education 136
community impact model 114
companionship, need of 109
compassion 12
compensatory mechanism 177
competition 17
compulsive behaviour 212
condoms 83, 123, 124
confusion 14, 15, 16, 17, 18, 20, 60, 125, 186, 188
   collective 67
congenital disease theories 1
congruency 11, 12, 14–15
contagion: heterosexual images of 122
contamination with past business
   see unfinished business
contextual situation, crisis resolution within 63
'contract' aspect 85–6, 124
   therapeutic 226
control group studies 166
convictions 151
coping potential 48, 52
coping strategies 51–2
   absence of 54
core identity gender 180
corruption notion 190
counselling 95, 160, 190
   attitude to 202
   lack of 184
counterculture 69–70
counter-transference 225–6
coupling 21, 25
crises other than 'coming out' 190–2
crisis 6, 35, 38–64
   AIDS *see* AIDS crisis
   coming out *see* coming out crisis

containment 52
definition xii, 42, 47
diagnosis 220
emotional prerequisites 40
growth-promoting concept *see*
    growth-promotion perspective
    of crisis
identity *see* identity crisis
identity growth and 165–205,
    207–8, 218
metaphysical interpretation 57–8
perspectives 56
phases 59–60
resolution 56
sequence of 48
sub-cultural 72–3
crisis clinics 219
crisis defusion in AIDS 133
crisis holding 51
crisis identification: AIDS and 127
crisis intervention 39–41, 47–64, 218–31
    AIDS 125–37
    and the critical stage 47–54
    fear of 45
    homosexual 55–64, 218–31
crisis moment 50, 51
crisis resolution 56, 62, 219
crisis responses 44–6, 170
crisis services 187
crisis theory 41–7
    antecedents to 38–41
critical stage 50–4
    identification 50–1, 52
cross-dressing 13, 14, 32, 58, 82
cross-meetings 93
'cruising' 52, 78, 80, 101 *bis*
'cruising spots' 105–10; *see also*
    meeting places
cues 101, 106–7, 215, 222
cultural activities 95
cultural displacement 67
cultural heritage 67
culture 66–8
    essential components 68
    popular 67–8
    procreation 10, 104
cumulative events 208
cytomegalo virus 135

dancing 102

daydreams 17
death wish 214
debilitating effects of coming out 183–4
defiance 139
demography 84
    in Greater Cape Town 90
denial 17, 45, 181, 210, 214
dependency needs 45
depression 184, 186, 231
    AIDS related 128
desire for same gender *see* same
    gender attraction
despair 187, 208, 211
determinism 38
developmental crises 43–4
developmental issues 62
deviancy 31
diagnosis 15, 135, 220
dialogue 97
difference, sense of 7–8, 11, 60–1, 171,
    184, 210
    and gay sub-culture 8, 72
    need to perpetuate 195
differences, perceived 193–6
discotheques 98–104, 110
discrimination 153, 177–8
    legislation 183, 218
dishonesties, past: reconciliation of 63
dispossession, cult of 102
dissolution of relationship 25
District Six 91
'disturbing' features of coming out
    crisis 187–8
diversity of sub-culture 73
domestic alliances 21
dominance conflicts 30
double-bind effect of sub-culture 71–2,
    84, 124–5, 167–8, 211–3, 231
drag 33
dreams 17
dual existence 178
duality of experiences 6, 57
Dutch Reformed Church 144–5

earlier conflicts, stress related to 63,
    208, 219
eating houses 102–3
economic inequalities, racial 145
ego, human 10
egocentric components of crisis 208

ego-destructive oppression 142
ego-dystonic disturbance 28
ego identity and sexual identity 8
ego state and crisis 40
ego theories 39–40
ejaculatory problems 16, 18–19
emotional aspect of homosexuality 5, 9
emotional flooding 134
emotional needs, denial of 80
emotional regulation, need for 81
emotions, denial of 45
employment discrimination 153, 177
encounters, public 96–7
End Conscription Campaign 94
enmeshment concept 9, 194–5
Equus 156
erection 18–19
    uncontrolled 16
Erikson's stages of identity
    development 3, 6, 7
erotic reading material 82
erotica 14, 81
eroticism 5
erotophilia 80
estrangement 24, 187, 215
exclusivity 215
exhibition in public 107, 150
existential authenticity 56
existential conflict 57
existential difference 60–1
Exit 104, 157, 203
experience: sociocentric features 61–2
experiences, duality of 6
experimental studies 165–6
experimentation, infantile 13
exploitation, sexual 190
exposure of genitals 107, 150

family 23–4
    acceptance 181
    involvement, lack of 24
    relationships 161, 186
family dynamics 10
family system 23–4
family systems, alternative 24, 194–5, 205
fantasies 7, 11
fantasies, homosexual 82, 208
    diagnosis and 222–3
    identity development and 11, 13,
        14–15, 16, 17, 18–19
    ownership 11, 13
    suppression of 23, 181
    testing of validity 83
fantasy apparatus 9, 58, 75
fantasy consolidation 17, 223
fantasy investigation (in crisis
    intervention) 221–5
fantasy system 5
fantasy types (chart) 224
father figure, need of 18
fear 13, 15, 20, 21, 26, 208
    AIDS related 128
    of exposure 73, 203
    of intervention 45
feedback 13, 23
fellatio 82, 147
feminism 27, 140
field dependent/independent 37
field independent 37, 195
field studies 165
field survey 168
fingers, use of 82
fondling 107
forced removals 93
Freud 1, 38–9
frigidity 16
fringe culture 69–70, 104, 109–10, 215
frustration 55
    AIDS related 128
functional disorganization 59, 61, 186
fundamentalism 2
future anticipation 62

gain and loss 48–50, 207
    AIDS related 129, 130, 136
gain potential 49, 207
gain through challenge, loss of 133
GASA 95, 169, 171, 197–200, 217
    counselling centres 190
    demise 158, 159
    International Gay Association and 140
    service activities 94–5, 156–9
    60-10 centre 93, 156, 158, 197, 204, 220
gay: definition xii
Gay Alliance of South Africa 159
Gay Association of South Africa *see* GASA

gay bashings 106
Gay Between 156
Gay Christian Movement 156
gay community 22
  definition xii–xiii
  see also gay movement, formal;
    sub-culture, homosexual
gay family notion 194, 205
gay identity acquisition 9
  see identity, homosexual for
    references to homosexual (or
    gay) identity
gay left 70
Gay Left Collective 140
gay liberation *see* liberation
gay movement, formal 138–64
  identity issues and 217–18
  in South Africa 154–60
  international perspective 138–42
gay priorities 158–9
gay rights 139, 232
  pursuit of 141
  see also discrimination; liberation
gay sub-culture *see* sub-culture,
    homosexual; gay movement
Gays Anonymous 156
gender
  coming to terms with 7
  definition 5
gender attributes 9
gender identity *see* identity, gender
gender inadequacy 8
gender role 180, 195–6
gender systems, external 10
genetic theories 1
genital area fixation 78
ghetto mentality 195
glossary 247–50
GLOW 157, 158, 217
going public 177, 178, 196
  compared with 'coming out' 178
  Golan's Stages in Crisis Intervention
    41–64 *passim*
Goodwood Show Grounds 176
Graaf's Pool 106
Greek homosexual sub-culture 65
grief, unresolved or delayed 60
Group Areas Act 91
growth
  AIDS and 214

growth potential 50, 51, 58–9, 219
growth promoting features of AIDS
    136, 213, 214
growth promotion capitalization 61,
    219
growth promotion perspective of crisis
    47, 167, 185, 219
guilt feelings 14, 15, 20, 23, 161, 184,
    186, 191
  connected with AIDS 127, 128

hazard, AIDS-related 150
hazardous event 57, 58–9
  AIDS 125–6
hazardous situation 57
health care, homophobic 192
health practitioners 196, 220
hedonistic value system 162
helping others 200–2
hermaphrodism 31
hero-worship figures 17
heterophobia 4
heterosexual encounters, fear of 186
heterosexual expectations, over-
    subscribing to 60
heterosexual homosexuality 33–4
heterosexual models 14
heterosexual panic 33–4
heterosexual perceptions of
    homosexuality 122
heterosexual relationships, severing of
    216
heterosexuality, presumed 58
Hillbrow 97
HIV 26, 101, 113
  'camping' linked with 109, 110
HIV-antibody positive 135, 136
HIV-antibody testing 136
holistic model 10
homo-erotic fantasies 11
homophile organizations 140, 196–7
  need for 211
homophobia AIDS and 119, 121, 124,
    125
  generalized 45, 161
  homosexual 73, 99, 108, 119, 193,
    199
  internalized 36, 63, 109, 187
  public 58, 141
homophobic myths 56

homophobic therapists 196, 204
homosexual behaviour, onset of 12
homosexual behaviours,
    institutionalized 21
homosexual crisis *see* crisis
homosexual experience 181
homosexual ideation 182
homosexual identity *see* identity,
    homosexual
homosexual liberation *see under*
    liberation
homosexual marriages 153
homosexual panic 28, 74–5
homosexual sensation 13, 180, 189, 210
homosexuality
    definitions xii, 4–6
    origin of term 1
    perceptions of 1–4, 65–6
homosocial people 215
hope 48, 52
    abandonment of 135
    during reintegration 62
    in AIDS 134, 136
human rights of homosexuals 139, 232
    pursuit of 141
    see also discrimination; liberation

iconography 70, 101, 110
idealization of homosexuality 62
identity
    definition 5, 6
    dichotomy 184
    owning of 8–9
    right to choose 73
identity, ego 8
identity, gender 22–3
identity, homosexual 181, 182
    AIDS and 119, 121–2, 123
    beginning achievement of 11, 17–19
    crisis and *see* crisis *and* identity
        crisis
    definition 181
    differentiated from sexual identity
        182
    growth *see* identity growth
    homosexual sub-culture and 65,
        68–75, 166–8, 169
    legal status and 141
    legal threats and 151
    oppression and inhibition of 142

    sexual exploration as affirmation of
        83
identity, sexual 1, 182
identity, social 7
    vs individual 188
identity, sub-cultural 7; *see also* sub-
    culture
identity acquisition 9
identity challenge 11, 14–16
identity commitment 11, 20–1
identity conflict 86
identity confusion *see* confusion
identity consolidation 6, 11, 21–7, 72
identity crisis 7, 20, 55–6, 207, 215
    diagnosis 220
    religion and 192
identity diffusion 11, 12–14
identity exploration 11, 16–17
identity feelings, negative 14
identity growth/development 1–37,
    166–7, 179–80, 189, 206–14
    crisis and 165–205, 207–8, 218; *see
        also* identity crisis
    pattern 210
    proposed model 10–27
    sexual 1–37, 180, 189, 206–14
    stages 11, 189, 210–11
    sub-culture and 166–8, 169
identity intolerance 20
identity patterns 7
ideological expression vs gender
    attraction 9
idiosyncratic behaviour and opinions
    166
idiosyncratic experience 7, 13
imagination 102
Immorality Act 94, 148–50, 154
immunological overload theory 114
impotency 16, 18, 23, 52, 108
incest, social 85, 103
incidence of homosexuality 84, 90, 91
incidents, cumulative, and identity
    growth 167
incorporating the other 80
independence vs dependence 60
individuation 12 *bis*
indoctrination, sub-cultural 78
infancy and early childhood 12–14
infantilism theories 1
inner reality 18

inner worth, sense of 14
inside-outside social network 95
insight, gaining of 61
integration of social institutions 94
interactive model (AIDS) 114
internal/external acceptance 3
internal dialogue 7, 15
internalization 12, 15
International Gay Association 140
interpersonal relationships, structure of 95
'intersexual' theories 2
intervention in crisis *see* crisis intervention
intimacy vs isolation 7
intimidation 13
inversion, homosexuality as 1
isolation 7, 8, 186, 187, 231

job discrimination 153, 177
Johannesburg Botanical Gardens 106
Johannesburg homosexual institutions 97, 98, 99, 100, 103
'junkie' syndrome 84

Kaposi's Sarcoma 112, 117, 135
Kinsey, A C 2

labelling 20, 22, 121, 227
Lago 157
Lambda 156
leadership 161–2, 217
learning theory concepts 10
leftist homosexuals 33
legal enactment/legal enforcement 153
legal freedom 141
legal identity, absence of 23
legal-physical oppression 142
legal position 146–54
　feelings about 36
legislation 117, 139–40
　apartheid 143
　application 150–4
　coming out and 183
　implementation 152
　interpretation 149–50, 152
　pro-homosexual 140
lesbian feminist movement 140
lesbianism 99
　political 27

lesbians 94, 97, 99–100
liberation, homosexual/gay 140, 141, 231–2
　homophile association and 197
　in South Africa 142–54, 217–18
　people liberation 95
liberation group 198
liberation movements 66, 138–54
　coming out and 183
liberation politics 141
life crisis, general 62–3
lifestyles: effect of AIDS 214
Link/Skakel 157
literature, gay 104–5, 139
lobby groups 161
loneliness 186, 198, 212
longitudinal process of identity formation 7
lores 22, 95
loss 44, 48–50, 185, 231
　acceptance and incorporation 61
　accumulated 132–3, 136
　anticipated 132–3
　camping out remedy 109
　collective 129
loss fixation 136, 207

macro-family 194
mainstream homosexual network 94
management of affect 61
marginality 55, 68, 104
marriage and homosexuality 153, 216
married men: camping 107
Marxism 27, 33, 94
masculine assertion 17
masculinity issues 191
masochistic fantasies, sado- 223
masochistic sex, sado- 82, 110
masturbation 14, 23, 82, 83, 101, 150
　period of stopping 17, 20
masturbatory-fantasy sequence 18
maturational crises 43–4
medical aid 179
meeting places 18, 71, 93 *bis*, 95–111
　monophile organizations 197, 198, 212
mental abnormality and homosexuality 5
mental disturbance 28
mental health practitioners 196, 220

mental illness, homosexuality listing as 140
meta-crisis 63, 72, 121, 213
meta-messages 9
meta-oppression 94
meta response 60
metaphysical approach to crisis 57–8
methodologies of research 165, 166–8
miasmic anti-homosexual bias 58
micro-culture 7
micro-family 194, 202
minority group status 195
mirror symbolism 95
missing, crisis of 58–9
mobility 176, 187, 212
'moffie' 73, 87 n, 93
moral judgements 2
moral opposition 161
moral panic 121
morality, sense of 13, 14, 16
moratorium 17, 20
mother: bonding and separation experiences 12, 13
mourning 182
multiple partner relationships 80
multiple relationships 25, 124, 125
Muslim attitude 163–4
mutuality in relationships 12

Napoleonic Code and homosexuality 65
Napoleonic era 1
narcissism 11, 17, 18, 66, 79, 171, 223
National Gay Convention 146
National Gay Task Force 140
nationalist culture 67, 144
negative defences 54
neurological factors of AIDS 116
new wave movement 70
nocturnal emissions 14, 17
normlessness 197
'nympholepsy' 78

object losses 44
objectification 107
obsessive disorders
   AIDS related 129
occupation 176–7
   choice of 177–8
   discrimination 153, 177

occupational-financial oppression 142
occupational freedom 178
offences 146–50
older men: discrimination against 174
OLGA 157, 217
oppression 94, 142, 195, 227
   and sub-culture 71
oral fixation 78
oral sex 82, 147
Organization of Lesbian and Gay Activists (OLGA) 94
ostracism xiii, 13; *see also* stigma
over-attachment 80–1
overcompensatory behaviour 46
overindulgent sexual behaviour: AIDS and 123

paedophile fantasies 223
paedophiles and homosexual behaviour 66, 190
parent/child relationship 207
parent objects, responses to 12–13
parental action 14
parental approval 13
parental disapproval 13, 62, 186
   fear of 13, 184
parental perceptions 13
parental rejection 186
parental support 24
parents, estrangement from 24
parents, responses of infant to 12–13
past conflicts 63, 208, 219
pathogenic crisis 208
peer group disapproval 13
penal codes 139–40
penile trauma 19
penis size 16
penis size obsession 78
percentage homosexualism 84, 90, 91
performance anxiety 214
personal integrity vs heterosexual values 57
personal model 98
personality profile, basic 229
phantom homosexual behaviour 16
phobias 17
pick-ups 107
'piece' 102
Pink Democrats 94, 157
Pink Triangle 94

Pneumocysties Carinii Pneumonia 112, 117, 135
police 153–4
　attitude to 152
police entrapment 150, 151
police harassment 151, 153
police raids 106
political activity 140, 213
political attitudes 94
political discourse on sexual behaviour 166
political ideologies 26, 158, 215
political issues 26–7
political reform and gay liberation 95
political splits 158
politics, homosexual 140
politics, liberation 141
pornography 81
positional model 98
power, testing out of 18
power concept 198
power conflict and homosexual behaviour 30
powerlessness, feelings of 20, 191, 197
precipitating factor 60–1
　AIDS related 130–2
　identification 221
　validity 131–2
predating, sexual 28
prejudice in work situation 178
premature ejaculation 16, 18
primitive fantasies 11
procreation
　homosexuality defies laws 65
　metaphorical 9, 71, 168
professional intervention *see* crisis intervention
progressive homosexuals 94
projection of crisis 193
promiscuity 83, 107, 211, 212, 231
　AIDS/HIV and 109
prosecutions 151
protocols 10, 18
pseudo-homosexuality 16, 28
pseudo-lesbianism 31
psychoanalysis 39
psychodynamic concepts 10
psychological battering, AIDS and 122
psychological birth 12
psychometric investigation 165

psychotherapy and coming out 190
puberty 11, 14–16, 210
public image manifestation 78
public leadership, lack of 161
public ownership 159–60
publications 82, 104–5, 156, 157, 165

questionnaire 241–6
　feelings about 202–4
　study 165–205

racial attitudes 94, 218
racial integration 94
radical gay activism 140
Radio 702 Crisis Clinic 219, 220
Rand Gay Organization 157
rape 16
　and coming out 190
　case histories 16, 51–2
rape trauma 51–2
reaction formation 17
reading material, erotic 82
reality, rehearsals for 61
reality base 11, 14, 15, 18, 24
recreation, sex as 107
regressive behaviour 40
regret and emotional growth 58
reintegration 50, 51
　AIDS related 134–6
　period of 62–4
rejection 161
　avoidance of 80, 108, 181, 186
　by parents 186
　gay clubs as response to 102
relationship behaviour 84–6
relationship construction 185
relationships 24–6, 231
　failure of 214
　resources for mutuality in 12
　urgency 25
religious factors 144–5, 161, 191–2, 204
religious groups 94, 95, 213
rent boys *see* sex workers
repressed behaviour 57
research design 166–8
research into homosexual issues 165–205
residential factors 175–6
residues 7, 38, 208; *see also* unfinished business

restaurants 103
retreat 45
retribution concept 162, 213
rewarding of homosexual behaviour 13
rights 139, 232
    pursuit of 141
    see also discrimination; liberation
risk
    AIDS 118–19, 125, 127, 129
    coming out 185
    crisis and 49, 53–4, 63
risk-oriented approach 52
ritual 9, 21, 95, 97, 101, 194, 195, 215
role crisis 43
role models 20, 211
role performance 66
'rough trade' 107, 110 n
rumours 122

sado-masochistic fantasies 223
sado-masochistic sex 82, 110
safer sex campaigns 136
safer sex guidelines 137
safer sex kits 130
safer sex practices 97, 107, 116, 123, 124, 132, 213, 214
same-gender attraction 13, 14, 15, 184
    acknowledgement of 15
    age of onset 170, 179, 180
    diagnosis 15
same-gender experimentation 210
sampling 169
Sandy Bay 106
saunas 95
schizophrenics 28
Scientific Humanitarian Committee 138
scientific revival 139
Sea Point 81, 106
searching 11, 17, 20, 49, 61, 78, 81, 83, 212
    AIDS related 129
security 111
seduction 16, 30, 190
Select Committee 155
self-acceptance 3, 11, 20, 181, 182
self-actualization 11
    sub-culture and 22
self-condemnation 132
self-control, loss of 48
self-defence 3–4

self-disclosure at work, AIDS scare and 178–9
self-esteem
    affirmation of 18
    AIDS and 122–3
    homosexual identity and 11, 14, 15, 34–6
    identity development and 11, 14, 15
    in crisis 61, 186, 208
    lack of 151, 184
    service organization and 196
    vulnerability 60
self-image, loss of 48
self-labelling 20, 22
self-obsessive behaviour 119–20
self-oppression 3, 26, 63, 71, 181
    legislation and 152
    sub-culture and 22
self-perception 12, 56, 188
self-recognition of homosexual identity 20, 182
self-verification and transitory sex 83
separateness, sense of 7, 8, 15, 58, 171, 215, 227
    sub-cultural 72
separation-individuation process 12
separatist concept of identity 72
separatist philosophy 3
service centre: attitudes towards 196–7
sex components, primacy of 211
sex partner preference 13, 77
sex role orientation 13
sex role preference 77
sex workers 29, 97
sexism 99
sexual addiction 80, 108, 212
sexual affirmation 211, 212, 231
sexual behaviour 75–84, 96–7
    AIDS risk factor 120–1
    dichotomies 77
    guides to 82
    identity development and 11, 13, 14–15
sexual commitment 21
sexual confusion 14
sexual deviance concept 9
sexual dysfunction, safer sex and 214
sexual encounter, first 182
sexual ethics, exposure to 14
sexual experimentation 213

sexual fixations 78
sexual identity see identity, sexual
sexual object of desire, composite 18
Sexual Offences Act (1988) 164
sexual orientation 8
  definition 166
sexual partner identity 180
sexual preference 8
sexuality 22–3
  and sub-culture 73
  definition 4
shame, sense of 14, 184, 186
shock, AIDS 127
signification 159
similarity of object to self 80
sin, homosexuality seen as 65
situational crisis 43, 44
situational homosexuality 28–9
60-10 organization see under GASA
'size queens' 78
social identity: development 7
social institutions, gay/homosexual
  admission 18
  Cape Town 95
    see also meeting places
social interaction, non-sexual 200
social network, gay: disadvantages 95
social oppression 142
social processes 101–2
social reform and gay liberation 95
social space of homosexual
  relationships 195
socialism 94
sociocentric features of crisis 61–2, 208
sociocentricity 191–2
sociological perspective 139
sociological theories 10
sodomy 65, 117, 146–7, 147, 150, 153
South African Airways 178, 204
South African Society 142–6
sport associations 95, 96
stages of homosexual identity
  development 11
state losses 44
statutory offences 148–50
stereotyping 76
stigma 46
  AIDS and 121, 122, 213
  and self-esteem 35, 36, 62
  homosexual sub-culture and 71, 95

internalized 36, 184
Stonewall riots 141
stress 23, 47
  identifying areas of 61
'stroking' 80, 107
stylization 110
sub-cultural acquisition 21
sub-cultural assimilation and rejection
  167
sub-cultural identity 7, 8
sub-cultural influence 9
sub-culture 68–9
  definition xiii, 69
sub-culture, alternative 70, 215
sub-culture, homosexual/gay 69–72,
  210–13
  AIDS and 117–23
  as surrogate family 194–6
  attitudes to 171, 212
  Cape Town, Greater 89–111
  code of conduct 25–6, 194
  coming out facilitated by 189
  construct of sexuality 78–9
  definition 69
  disengagement from 195
  diversity of 73
  fear of exposure to 57, 120
  history 65–8
  homosexual identity and 8, 65–88
  interaction with 18, 21–2
  liberation and 162
  love-hate relationship with 73–5,
    108
  peripheral 17
  rejection of 211–12
  responding to 57
  sense of difference and 8, 72
  separateness fostered by 8, 72
  state of crisis 72–3
sub-species concept 65
suicidal thoughts and attempts 120,
  132–3, 184, 185, 186, 187
super-ego 14
suppressed material, dealing with 56–7
suppression of fantasies 23
surrogate family 194–5
symbolic content of identity 166
symbolic interaction approaches 6
symbolism 5
  mirror 95

symbols, gay 70, 215

task forces, gay 141
tax rights 153
territorial imperative 103, 111 n
territory of ownership 101
testing out behaviour 17, 18, 83, 98, 106
Thanatos 213, 214
theories of homosexuality 1–4
therapeutic relationship 226–8
therapists
   attitude to homosexual issues 196, 204, 226–8
   homophobic 196, 204
third sex concept 33, 65
thought patterns, development of 61
threat 59
titles of participants 85
tongue, use of 82
tourists, gay 93
transference and counter-transference 225–6
transience 80, 104, 108, 199, 211; *see also* camping; casual sex
transient homosexuality 28
transitional crisis 43
transitional object syndrome 79–81, 88 n, 199, 214
transsexuality 31–2
transsexuals 97
transvestism 32
transvestites 97
trauma, coming out 185
traumatic incidents 16
traumatic pitch 17
traumatic sexual experience
   bearing on sexual identity 190
trial and error behaviour 49

tricameral parliament 143
type 18

Ulrich, Carl 138
unfinished business 23, 61, 208, 226
   AIDS and 127, 209
   see also residues
Unite 156
United States gay movement 140, 187
unnatural offences 147, 150
Uranian 1
urbanization 175
urgency in active crisis 62
urgency of bonding 25

validity of crisis experience 54, 61
validity of identity
   confirmation of 83
   denial of 56
venereal diseases, exposure to 124
vernacular 78–9, 85, 215, 247–50
victimization 195
   inverse 21
visibility of homosexual institutions 98–9
visibility of homosexuals 84
vocational planning 178
vulnerable state 59–60, 202, 210, 231
   AIDS 126–30

weakness, sense of 191, 208
wet dreams *see* nocturnal emissions
white gays: racial attitude 94
withdrawal 45, 58, 60, 184, 186, 187
work paranoia 178
World Council of Churches 140
World Health Organization 140
world view 95
worried well 116, 117, 118–19, 135